Studies in Big Data

Volume 194

Series Editor

Janusz Kacprzyk, Polish Academy of Sciences, Warsaw, Poland

The series "Studies in Big Data" (SBD) publishes new developments and advances in the various areas of Big Data- quickly and with a high quality. The intent is to cover the theory, research, development, and applications of Big Data, as embedded in the fields of engineering, computer science, physics, economics and life sciences. The books of the series refer to the analysis and understanding of large, complex, and/or distributed data sets generated from recent digital sources coming from sensors or other physical instruments as well as simulations, crowd sourcing, social networks or other internet transactions, such as emails or video click streams and other. The series contains monographs, lecture notes and edited volumes in Big Data spanning the areas of computational intelligence including neural networks, evolutionary computation, soft computing, fuzzy systems, as well as artificial intelligence, data mining, modern statistics and Operations research, as well as self-organizing systems. Of particular value to both the contributors and the readership are the short publication timeframe and the world-wide distribution, which enable both wide and rapid dissemination of research output.

The books of this series are reviewed in a single blind peer review process.

Indexed by SCOPUS, EI Compendex, SCIMAGO and zbMATH.

All books published in the series are submitted for consideration in Web of Science.

Simeon Okechukwu Ajakwe · Dong-Seong Kim
Editors

Blockchain for Security of Autonomous Intelligent Vehicles

Cyber-Cognition Mechanisms for the Security of Autonomous Intelligent Vehicles (AIS)

Springer

Editors
Simeon Okechukwu Ajakwe
ICT Convergence Research Centre (ICT-CRC)
Kumoh National Institute of Technology
Gumi, Korea (Republic of)

Dong-Seong Kim
Networked Systems Laboratory, Department of ICT Convergence Engineering
Kumoh National Institute of Technology
Gumi, Korea (Republic of)

ISSN 2197-6503 ISSN 2197-6511 (electronic)
Studies in Big Data
ISBN 978-3-032-20850-7 ISBN 978-3-032-20851-4 (eBook)
https://doi.org/10.1007/978-3-032-20851-4

This Springer imprint is published by the registered company Springer Nature Switzerland AG
The registered company address is: Gewerbestrasse 11, 6330 Cham, Switzerland

To the Giver of life and Upholder of all things, I remain eternally grateful.

To the dreamers who look at a horizon filled with intelligent machines and see not risk, but possibility.

To the innovators who believe that trust, security, and transparency are not limitations, but the foundations of progress.

To the researchers and engineers who devote their minds and hearts to building autonomous systems that safeguard every life they touch.

To the guardians of ethical technology, who remind us that the future must not only be smarter, but also safer, fairer, and more human.

And above all, to those who envision a world where intelligence moves freely, responsibly, and securely through every road and sky, this book is inspired by your courage and dedicated to your unwavering pursuit of a better tomorrow.

Foreword

The rapid advancement of Autonomous Intelligent Vehicles (AIVs) is reshaping modern mobility through the convergence of artificial intelligence, pervasive sensing, and high-speed V2X communication. As these systems transition from research prototypes to large-scale deployments, the need for secure, trustworthy, and resilient communication frameworks becomes paramount. Traditional centralized security models, particularly PKI-based credential systems, struggle to meet the stringent requirements of low latency, high mobility, and adversarial resistance inherent in vehicular environments, as outlined in the foundational discussion of AIV security challenges.

Blockchain technology offers a compelling alternative by providing decentralized trust, immutability, and transparent auditability. Across the chapters in this volume, the authors demonstrate how blockchain can strengthen identity management, event logging, cooperative decision-making, and automated service execution through smart contracts. The architectural models, ranging from blockchain-enabled V2X networks to hierarchical UAV trust frameworks, highlight the versatility of distributed ledgers in securing both ground and aerial autonomous systems.

At the same time, the book confronts blockchain's inherent limitations, particularly the tension between security, speed, and scalability. This blockchain trilemma poses unique challenges in safety-critical mobility ecosystems that depend on millisecond-level responsiveness. The exploration of hybrid consensus mechanisms, edge-assisted blockchain layers, lightweight cryptography, and AI-augmented security provides practical pathways for reconciling these constraints, as examined in detail in the chapter on trilemma-oriented design for AIV ecosystems.

Equally notable is the treatment of smart contracts as governance tools for autonomous mobility. Their applications in automated tolling, energy trading, insurance, intrusion detection, and compliance verification are balanced by rigorous analyses of vulnerabilities and emerging quantum-era threats, underscoring the need for continual evolution in secure contract design.

This book offers a timely and well-structured contribution to the growing body of research on secure autonomous transportation. By integrating theoretical foundations, algorithmic advances, architectural models, and practical case studies, it provides a comprehensive guide for researchers, engineers, and policymakers seeking to build trustworthy AIV systems.

It is my pleasure to commend this volume to readers and to the broader scientific community committed to advancing secure, intelligent mobility.

Professor Jae Min Lee
Director, Department of Defense Studies
KICS
Gyeongsangbuk-do, Gumi, South Korea

Preface

The rapid rise of Autonomous Intelligent Vehicles (AIVs) marks a transformative moment in the evolution of modern mobility. From self-driving ground vehicles to autonomous aerial systems, the convergence of artificial intelligence, distributed sensing, and high-speed communication has reshaped how intelligent systems perceive, decide, and collaborate. Yet, as these autonomous agents assume greater control over safety-critical tasks, the need for trustworthy, transparent, and resilient computational infrastructures becomes essential. The complexities of securing such interconnected ecosystems reveal profound challenges that demand innovative, interdisciplinary solutions. This book was conceived in response to those challenges.

Spanning three cohesive parts, the volume explores how **blockchain technologies, when thoughtfully designed and integrated, can serve as the backbone of trust** for next-generation autonomous mobility. Each chapter contributes to a unified narrative that advances the foundations, architecture, security mechanisms, and governance structures required to support safe, scalable, and intelligent autonomous systems.

First Part provides the foundations for understanding the dynamic interplay between blockchain and AIV ecosystems. Chapters "Autonomous Intelligent Vehicles: An Overview," "Introduction to Blockchain and Autonomous Intelligent Vehicles," and "Blockchain Fundamentals in Autonomous Intelligent Vehicle" introduce readers to the fundamental characteristics of autonomous vehicles, the vulnerabilities they face, and the principles of distributed ledgers, consensus, identity management, and cryptographic mechanisms. These chapters establish the technical and conceptual groundwork upon which later contributions build—highlighting why traditional, centralized trust models falter in environments demanding low latency, high mobility, and strong resilience.

Second Part advances the discussion to architecture and intelligent frameworks, exploring how blockchain supports multi-agent systems and AI-augmented autonomous operations. Chapter "Blockchain in AIV with Multiple Agents AI Architecture" presents a comprehensive architecture for AIVs operating within complex, multi-agent environments. By integrating blockchain-enabled registries, decentralized identity systems, mixture-of-experts (MoE) models, and inter-domain mobility protocols, this chapter illustrates how distributed intelligence can enhance coordination, transparency, and accountability across heterogeneous agents. Chapter "Blockchain-Augmented AI- Empowered Autonomous UAV Security Systems: Towards Trustworthy Airspace Operations on the Internet of Drones (IoD)" extends this perspective into the aerial domain, demonstrating how blockchain and AI can jointly secure the Internet of Drones (IoD). It examines hierarchical trust systems, UAV authentication, privacy-preserving verification, edge-zone consensus, and cyber-resilient UAV operations—offering a unified framework for reliable autonomy across both ground and airspaces. Together, the chapters in the Second Part showcase how blockchain reshapes the operational landscape of intelligent mobility by enabling trustworthy coordination among autonomous vehicles, drones, and supporting infrastructures.

Third Part addresses the core security, governance, and scalability challenges facing blockchain-enabled AIV systems. Chapter "Threats and Security Challenges in Autonomous Intelligent Vehicles" delivers a detailed analysis of security threats targeting autonomous vehicles—from denial-of-service attacks and sensor spoofing to privacy breaches and communication vulnerabilities. It evaluates both classical cybersecurity measures and blockchain-based mitigation strategies, providing essential insight into safeguarding safety-critical autonomous systems. Chapter "Smart Contracts for Secure Autonomous Vehicle Operation" focuses on smart contracts as the programmable governance layer of AIV ecosystems. This chapter examines smart contract vulnerabilities, quantum-era risks, verification models, and automated trust mechanisms. It also presents a hybrid quantum–transformer learning approach for vulnerability detection, illustrating how emerging intelligence models can protect blockchain-based autonomous operations. Chapter "Blockchain Trilemma in Autonomous Intelligent Vehicle" concludes the volume with an in-depth exploration of the **blockchain trilemma**, the tension between security, scalability, and speed. The chapter introduces hybrid consensus models, edge-assisted blockchain layers, lightweight cryptography, and AI-driven validator governance to address the real-time performance requirements of autonomous mobility. It further presents case studies showing how tiered blockchain architectures can support both dense urban traffic and high-speed highway maneuvers without sacrificing integrity or responsiveness. Together, the chapters in the Third Part form a comprehensive guide to the threat landscape, automated operational governance, and performance constraints that define the future of secure, blockchain-assisted autonomous systems.

This book, structured across foundational theory, advanced architectures, and rigorous security frameworks, aims to serve as both a scholarly reference and a forward-looking roadmap. We hope it will inspire researchers, practitioners, and policymakers to rethink trust in autonomous mobility and to cultivate ecosystems where intelligence moves freely, safely, and transparently. By weaving together blockchain, AI, cryptography, security engineering, and mobility science, this volume offers a multidimensional perspective on one of the most critical technological frontiers of our time.

We extend our appreciation to the contributing authors for their scholarship, to our institutions for their support, and to the funding bodies whose commitment to research excellence made this work possible. Above all, we thank the global research community whose continued efforts advance the vision of secure, interoperable, and intelligent autonomous transportation.

Gumi, South Korea

Simeon Okechukwu Ajakwe
Dong-Seong Kim

Acknowledgements

The editors extend their sincere appreciation to all chapter contributors whose technical expertise, research dedication, and collaborative spirit have shaped this volume. Their efforts, spanning the disciplines of blockchain engineering, autonomous systems, cybersecurity, artificial intelligence, and intelligent transportation, collectively form the foundation of this book.

We gratefully acknowledge the institutional support that made this work possible. In particular, we recognize the ICT Convergence Research Center and NSLab Inc. at Kumoh National Institute of Technology for their continued leadership in research on autonomous intelligent vehicles and secure communication systems. Their laboratories, resources, and collaborative research environment strengthened many of the studies presented in this volume.

This work was supported in part by the **Ministry of Education, Science and Technology (MEST)**, the **Grand ICT Research Center Program**, and the **National Research Foundation (NRF) of Korea**. Their funding has been instrumental in advancing the research frameworks, experimental systems, and innovative methodologies that underpin several chapters of this book. We gratefully acknowledge their ongoing commitment to fostering excellence in science, technology, and future mobility research.

We also extend our gratitude to the reviewers and subject-matter experts whose insights and careful evaluations strengthened the clarity, rigor, and cohesion of this volume. Their constructive feedback significantly enhanced the book's scholarly contributions.

Our heartfelt thanks go to the editorial and production teams at Springer Nature for their professionalism, guidance, and support throughout the publication process. Their attention to detail and dedication to academic excellence ensured a seamless transition from manuscript to final volume.

Finally, we express our personal appreciation to our families, colleagues, and students, whose encouragement, patience, and steadfast support sustained us throughout this undertaking. As autonomous mobility continues to evolve, we hope that this book serves as a valuable resource for researchers and practitioners working to build secure, trustworthy, and intelligent transportation ecosystems for the future.

Glossary of Key Terms

Adversarial Attack A manipulation technique in which malicious inputs (such as altered sensor readings or perturbed visual data) deceive AI models, causing incorrect perception or decision-making in autonomous systems.

Aggregated Zero-Knowledge Proof (ZKP) A cryptographic mechanism allowing multiple proofs to be combined and verified efficiently without revealing underlying data—useful for privacy-preserving authentication in vehicle platoons and IoD systems.

Artificial Intelligence (AI) A computational discipline enabling machines to learn, reason, and make decisions. In AIVs, AI supports perception, prediction, path-planning, threat detection, and autonomous control.

Autonomous Intelligent Vehicle (AIV) A self-driving system capable of perceiving its environment, making decisions, and operating with minimal human intervention through integrated AI, sensor fusion, networking, and control logic.

Blockchain A distributed, tamper-evident ledger that records transactions across decentralized nodes. Used in AIV ecosystems to ensure trust, security, transparency, and immutability of communication records.

Blockchain Trilemma A fundamental constraint stating that decentralized systems cannot maximize **security**, **scalability**, and **speed/latency** simultaneously. AIV networks face this challenge acutely due to real-time safety requirements.

Byzantine Fault Tolerance (BFT) A consensus property that ensures system reliability even when some nodes act maliciously or unpredictably—critical for securing large-scale vehicular and drone networks.

Consensus Mechanism A protocol enabling distributed nodes in a blockchain network to agree on the validity of transactions. Examples include Proof of Work (PoW), Proof of Stake (PoS), and hybrid models for AIVs.

Cyber-Physical System (CPS) A system integrating computation, networking, and physical processes. AIVs are CPSs where real-time software decisions directly affect physical movement and safety.

Decentralized Identifier (DID) A self-sovereign identity construct used in blockchain-based AIV systems that allows vehicles or agents to authenticate securely without centralized authorities.

Denial of Service (DoS) Attack A cyberattack that overwhelms an AIV's network or computational resources, preventing normal operation and potentially leading to safety hazards.

Distributed Ledger Technology (DLT) A broader category encompassing blockchain and other decentralized record-keeping systems used to ensure integrity, transparency, and availability of vehicular data.

Domain-Based Registry A blockchain-compatible architecture that organizes vehicle or drone identities based on operational domains (e.g., highway, airspace), enabling scalable multi-agent coordination.

Edge Computing A computational model in which processing occurs close to the source of data—essential for reducing latency in AIV and IoD operations.

Fault-Tolerant Quantum Computing (FTQC) Advanced quantum computing capable of reliable, error-corrected computation. FTQC is relevant for evaluating future threats to classical cryptographic systems.

Flying Ad Hoc Network (FANET) A dynamic communication network of UAVs that self-configure and coordinate in real time, often secured using blockchain and AI frameworks.

Intelligent Transportation System (ITS) An integrated infrastructure using digital communication, smart sensors, and automation to improve road safety, mobility, and traffic efficiency.

Internet of Drones (IoD) A network architecture enabling cooperative, multi-drone operations with secure communication, trust management, routing, and airspace coordination—often supported by blockchain layers.

Internet of Vehicles (IoV) An ecosystem of interconnected vehicles, infrastructure, cloud services, and sensors facilitating V2V, V2I, and V2X communication for autonomous mobility.

LiDAR A remote sensing method that uses laser pulses to create 3D maps of surroundings, enabling accurate environmental perception for AIVs.

Mixture of Experts (MoE) Model An AI architecture that selects specialized "expert" models within a neural network to handle specific tasks—used in Chapter "Blockchain in AIV with Multiple Agents AI Architecture" for multi-agent AIV intelligence.

Multi-Agent System (MAS) A network of autonomous agents (vehicles, drones, roadside units) that interact, coordinate, and make collective decisions, often secured by blockchain-based identity and trust frameworks.

Post-Quantum Cryptography (PQC) Next-generation cryptographic algorithms designed to resist quantum-computer attacks. PQC is critical in AIV security as emerging quantum systems threaten classical cryptography.

Proof of Stake (PoS) A consensus mechanism validating new blocks based on node stake rather than computational power—often adapted for low-latency vehicular blockchains.

Proof of Reputation (PoR) A consensus model leveraging node reputation scores to validate blocks in decentralized systems, improving scalability and trust in vehicular networks.

Public Key Infrastructure (PKI) A centralized trust framework using certificates for authentication. Traditional PKI is limited in highly mobile, real-time AIV networks due to latency and revocation challenges.

Roadside Unit (RSU) A fixed communication node used in V2X networks to facilitate data exchange between vehicles and infrastructure.

Sensor Spoofing An attack in which false signals are injected into LiDAR, radar, or camera systems, causing the AIV to misinterpret its surroundings.

Smart Contract A self-executing blockchain program that automates agreements, transactions, or policies within AIV ecosystems. Used for tolling, insurance, updates, fleet coordination, and more.

Sybil Attack A security attack where a malicious agent creates multiple forged identities to manipulate a vehicular or blockchain network.

Vehicle-to-Everything (V2X) A suite of communication protocols enabling autonomous vehicles to exchange information with infrastructure (V2I), other vehicles (V2V), networks (V2N), and pedestrians (V2P).

Vehicular Ad Hoc Network (VANET) A decentralized mobile network of vehicles enabling direct communication for safety, routing, and situational awareness.

Zero-Knowledge Proof (ZKP) A cryptographic protocol allowing one party to prove the validity of information without revealing the information itself—used in Chapter "Blockchain in AIV with Multiple Agents AI Architecture" for privacy-preserving authentication.

Contents

Security Threats, Cyber Risks, Smart Contracts, and Blockchain Governance for AIV

Contributors

JohnPaul Adimonyemma Department of Industrial and Manufacturing Engineering, FAMU-FSU College of Engineering, Florida Agricultural and Mechanical University, Tallahassee, FL, USA

Chukwudi E. Agbaraji Department of Computer Science, Bishop's University, Sherbrooke, QC, Canada

George Chidera Akor IT-Convergence Engineering, Kumoh National Institute of Technology, Gumi, South Korea

Rubina Akter Department of Electrical and Software Engineering, University of Calgary, Calgary, Canada

Dan Ali School of Science, Engineering and Environment, The University of Salford, Salford, Greater Manchester, UK

Chimeremma Sandra Amadi Department of IT Convergence Engineering, Kumoh National Institute of Technology, Gumi, South Korea;
Department of Information Technology, Federal University of Technology, Owerri, Nigeria

Ikechukwu Ignatius Ayogu Department of Computer Science, Federal University of Technology, Owerri, Imo State, Nigeria

Chidi Ukamaka Betrand Department of Computer Science, Federal University of Technology, Owerri, Imo State, Nigeria

Lucky Uchenna Chibuike Directorate of General Studies, University of Agriculture and Environmental Sciences, Umuagwo, Imo State, Nigeria

Diovu Chidiebere Biomedical Engineering Department, Federal University of Allied Health Sciences, Enugu, Enugu, Nigeria

Harrison Osasogie Edokpolor University of Bath, Bath, UK

Victor Eghujovbo Mechanical, Automotive and Material Science, University of Windsor, Windsor, ON, Canada

Elvin Eziama Mechanical, Automotive and Material Science, University of Windsor, Windsor, ON, Canada

Mohtasin Golam NSLab Inc. ICT Convergence Research Center, Kumoh National Institute of Technology, Gumi, South Korea

Ugboaja Samuel Gregory Department of Computer Science, Michael Okpara University of Agriculture, Umuahia, Abia, Nigeria

Benedict O. Ibe School of Science, Engineering and Environment, The University of Salford, Salford, Greater Manchester, UK

Taesoo Jun Department of IT Convergence Engineering, Kumoh National Institute of Technology, Gumi, South Korea

Victor Ikenna Kanu IT-Convergence Engineering, Kumoh National Institute of Technology, Gumi, South Korea

Douglas Allswell Kelechi Department of Computer Science, Federal University of Technology, Owerri, Imo State, Nigeria

Ukabuiro Ikenna Kelechi Department of Computer Science, Abia State University, Abia State, Uturu, Nigeria

Dong-Seong Kim IT-Convergence Engineering, Kumoh National Institute of Technology, Gumi, South Korea;
NSLab Co. Ltd, Gumi, South Korea;
NSLab Inc., ICT Convergence Research Center, Gumi, South Korea

Jae Min Lee IT-Convergence Engineering, Kumoh National Institute of Technology, Gumi, South Korea

Ifeanyi F. Nwosu School of Science, Engineering and Environment, The University of Salford, Salford, Greater Manchester, UK

David A. Oduro Senior FP&A Analyst, Gulf of Maine Research Institute, Portland, ME, USA

Collins Izuchukwu Okafor IT-Convergence Engineering, Kumoh National Institute of Technology, Gumi, South Korea

Ogechi O. Onyebuchibe School of Science, Engineering and Environment, The University of Salford, Salford, Greater Manchester, UK

Christian Nduka Onyeukwu Department of Computer Science, Bishop's University, Sherbrooke, QC, Canada

Vadym Slyusar Central Research Institute of Armaments and Military Equipment of Armed Forces of Ukraine, Kyiv, Ukraine

Esmot Ara Tuli ICT Convergence Research Center, Kumoh National Institute of Technology, Gumi, South Korea

Ajere Ikenna Uzoma Department of Computer Science, Federal University of Technology, Owerri, Imo State, Nigeria;
University of Agriculture and Environmental Sciences, Umuagwo, Imo State, Nigeria;
Department of Information Technology, Federal University of Technology, Owerri, Imo State, Nigeria

Abbreviations

AEB	Autonomous Emergency Braking (if referenced in context)
AI	Artificial Intelligence
AIoT	Artificial Intelligence of Things
AIV	Autonomous Intelligent Vehicle
AUV	Autonomous Underwater Vehicle
AV	Autonomous Vehicle
AVs	Autonomous Vehicles
B5G	Beyond 5G (used in future mobility discussions)
BFT	Byzantine Fault Tolerance
BLE	Bluetooth Low Energy (appears in V2X contexts)
BTS	Base Transceiver Station (implied within IoD architectures)
CAN	Controller Area Network
CNN	Convolutional Neural Network
CPS	Cyber-Physical System
C-V2X	Cellular Vehicle-to-Everything
CVIS	Cooperative Vehicle-Infrastructure System
DAG	Directed Acyclic Graph
DDoS	Distributed Denial of Service
DID	Decentralized Identifier
DLT	Distributed Ledger Technology
DoS	Denial of Service
E2E	End-to-End
ECU	Electronic Control Unit
ETSI	European Telecommunications Standards Institute
F1 Score	Classification Accuracy Metric
FANET	Flying Ad Hoc Network
FTQC	Fault-Tolerant Quantum Computing
GNSS	Global Navigation Satellite System
GPS	Global Positioning System
GUTI	Globally Unique Temporary Identifier (telecom context)
HIL	Hardware-in-the-loop (contextual in testing frameworks)

IoD	Internet of Drones
IoV	Internet of Vehicles
ITS	Intelligent Transportation System
IV	Intelligent Vehicle
IVTP	Intelligent Vehicle Trust Point
KEM	Key Encapsulation Mechanism (Post-Quantum Cryptography)
LiDAR	Light Detection and Ranging
MAC	Medium Access Control (communications layer)
MANET	Mobile Ad Hoc Network
MIL	Model-in-the-loop
ML	Machine Learning
ML-DSA	Module-Lattice Digital Signature Algorithm (PQC)
ML-KEM	Module-Lattice Key Encapsulation Mechanism (PQC)
MoE	Mixture of Experts
NISQ	Noisy Intermediate-Scale Quantum
NIST	National Institute of Standards and Technology
NTN	Non-Terrestrial Networks
OTA	Over-the-Air (Updates/Communication)
PKI	Public Key Infrastructure
PoR	Proof of Reputation
PoS	Proof of Stake
PoW	Proof of Work
PQC	Post-Quantum Cryptography
QKD	Quantum Key Distribution
QLAN	Quantum Local Area Network
QMAN	Quantum Metropolitan Area Network
QNN	Quantum Neural Network
QWAN	Quantum Wide Area Network
RE	Rear-End Collision (classification label in Chapter "Blockchain Trilemma in Autonomous Intelligent Vehicle" examples)
RSU	Roadside Unit
SAE	Society of Automotive Engineers
SLH-DSA	Hash-Based Digital Signature Algorithm (PQC)
SPoF	Single Point of Failure
SSID	Service Set Identifier
STARK	Scalable Transparent Argument of Knowledge (ZK-proof system)
Sybil Attack	Identity-Forging Attack Model
TD	T-bone Collision (classification label)
TDMA	Time Division Multiple Access
TLS	Transport Layer Security
TPM	Trusted Platform Module (implied in hardware-trust contexts)
UAES	University of Agriculture and Environmental Sciences
UAV	Unmanned Aerial Vehicle
UE	User Equipment
V2I	Vehicle-to-Infrastructure

V2N	Vehicle-to-Network
V2P	Vehicle-to-Pedestrian
V2V	Vehicle-to-Vehicle
V2X	Vehicle-to-Everything
VANET	Vehicular Ad Hoc Network
WAVE	Wireless Access in Vehicular Environments
Wi-Fi	Wireless Fidelity (802.11 standards)
ZKP	Zero-Knowledge Proof

Foundations of AIV and Blockchain Technologies

Autonomous Intelligent Vehicles: An Overview

Collins Izuchukwu Okafor, George Chidera Akor, Victor Ikenna Kanu, Jae Min Lee, and Dong-Seong Kim

Abstract Autonomous Intelligent Vehicles (AIVs) are rapidly evolving from standalone systems into complex, interconnected nodes within vast cyber-physical ecosystems. This chapter provides a foundational overview of the AIV landscape, establishing the context for the critical security challenges addressed throughout this book. We classify AIVs across terrestrial, aerial, and maritime domains and examine the core technologies that enable their autonomy, including AI-driven perception, IoT-based sensing, edge computing, and 5G/6G communications. By analyzing the system architecture and communication networks (e.g., V2X), we illuminate the inherent, multi-layered vulnerabilities that traditional security models fail to adequately address. We explicitly frame blockchain as a *security enabler* for AIVs by linking key vulnerabilities (LiDAR/GNSS spoofing, C-V2X DoS, data or identity forgery) to ledger-based defenses: tamper-evident log, decentralized identity, and auditable smart contracts, and by illustrating, through comparative tables and scenario-driven designs (platooning, UAV swarms, maritime cargo), how permissioned blockchains strengthen standards-driven security in practice. This overview

Collins Izuchukwu Okafor, George Chidera Akor, Victor Ikenna Kanu, Jae Min Lee and Dong-Seong Kim—These authors contributed equally to this work.

C. I. Okafor (✉) · G. C. Akor · V. I. Kanu · J. M. Lee · D.-S. Kim
IT-Convergence Engineering, Kumoh National Institute of Technology, Gumi, South Korea
e-mail: collinsokafor@kumoh.ac.kr

G. C. Akor
e-mail: georgeakor@kumoh.ac.kr

V. I. Kanu
e-mail: vkanu@kumoh.ac.kr

J. M. Lee
e-mail: ljmpaul@kumoh.ac.kr

D.-S. Kim
e-mail: dskim@kumoh.ac.kr

D.-S. Kim
NSLab Co. Ltd, Gumi, South Korea

Kumoh National Institute of Technology, Gumi, South Korea

S. O. Ajakwe and D.-S. Kim (eds.), *Blockchain for Security of Autonomous Intelligent Vehicles*, Studies in Big Data 194,
https://doi.org/10.1007/978-3-032-20851-4_1

serves as a crucial primer, demonstrating why decentralized, trust-based solutions like blockchain are not merely advantageous but essential for the secure and reliable future of autonomous mobility.

Keywords Autonomous intelligent vehicles · Communication networks · Drone networks · Edge computing · 5G/6G

1 Introduction

Autonomous Intelligent Vehicles (AIVs) now operate across land, air, and sea. They sense, decide, and act with minimal human oversight, and do so while connected to other vehicles, roadside units, cloud services, and sometimes satellites. Two landmark demonstrations set much of the technical agenda: the 2005 Defence Advanced Research Projects Agency (DARPA) Grand Challenge, where "Stanley" completed 212 km of off-road driving, and the 2007 DARPA Urban Challenge, where "Boss" negotiated urban traffic rules at road speeds. These projects codified the now standard pipeline of perception, localisation, motion planning and control, and they demonstrated that robust autonomy is possible under resource limits and environmental uncertainty [1, 2].

The modern AIV landscape is broader than passenger cars, as popularised by companies like Waymo and Tesla. It includes Unmanned Aerial Vehicles (UAVs), Autonomous Underwater Vehicles (AUVs) and purpose-built machines for logistics and mining. For example, recent AUV fieldwork used interferometric side-scan sonar to monitor aquaculture infrastructure in a high-energy, shallow-water environment, illustrating how autonomy supports inspection tasks that are costly or risky for divers [3].

Progress has been driven by two converging trends. First, on-board sensing and learning have improved markedly. Light Detection and Ranging (LiDAR), radar and multi-camera rigs feed deep models that now track, segment and forecast with state-of-the-art accuracy [4]. Second, vehicles increasingly cooperate with each other and with the road-side. Vehicle to Everything (V2X) connectivity underpins Cooperative Vehicle-Infrastructure Systems (CVIS), which extend situational awareness beyond line of sight, improve intent prediction and support earlier, smoother decision-making. Recent surveys synthesise this evidence and outline open issues, including how to deal with delays, packet loss and heterogeneity across the fleet [5].

Cooperative perception is a focal example. Sharing features or objects between vehicles and Roadside Units (RSUs) can raise detection rates, reduce false negatives and speed up hazard response, provided calibration and synchronisation are handled correctly. New datasets and benchmarks, such as TUMTraf V2X at the Conference on Computer Vision and Pattern Recognition (CVPR) 2024, are enabling more rigorous comparisons of methods and failure modes [6].

Despite these advances, safety and security remain central concerns. The attack surface spans sensors, communications and decision logic. Recent work shows that carefully timed LiDAR spoofing can remove pedestrians from detections even when the target vehicle is moving at typical urban speeds, which underscores the need for physical-layer defences and cross-sensor consistency checks [7]. At the communications layer, low-duty-cycle denial-of-service on Cellular Vehicle to Everything (C-V2X) sidelink scheduling can degrade channel availability enough to increase crash risk in studies that couple network loss to vehicle dynamics. Global Navigation Satellite System (GNSS) jamming and spoofing remain prevalent, which raises integrity questions for any stack that leans heavily on satellite fixes without cross-checking inertial and map priors [8–11].

AIV systems therefore blend multiple assurance measures. At the algorithmic level, redundancy and diversity mitigate correlated errors across sensors and models. At the system level, V2X links add complementary views, but those same links introduce fresh trust questions like "Who is allowed to publish safety-critical observations?", "how are identities rotated to preserve privacy?", and "how are misbehaving nodes excluded quickly without central bottlenecks that fail closed?" Recent tutorial reviews in wireless access and cooperative perception document the performance and privacy trade-offs involved in these choices and call out gaps in interoperability testing [12, 13]

Emerging designs propose lightweight consensus for UAV swarms and dynamic vehicular clusters, with attention to low-power nodes and intermittent links [13, 14]. Recent literature examines how blockchain can support V2X identity management, event logging and software-update provenance, while also highlighting latency, storage and privacy pitfalls that must be engineered around [15, 16]. As a result, in this chapter, we explore decentralised trust mechanisms, particularly blockchain, to provide tamper-evident logging, distributed authentication and cross-domain interoperability to temper the risks of a single administrative chokepoint found in classical Public Key Infrastructure (PKI).

Multi-access Edge Computing (MEC) is the other enabling thread. Many perception and prediction tasks are time-critical and bandwidth-heavy, which argues for keeping most inference on the vehicle while offloading selected summarised products to the network. Recent studies show that end-to-end latency budgets, packet loss and feature-level fusion quality jointly determine the safety benefits of cooperative perception, and that edge aggregation can help at complex junctions [17, 18].

Two principles guide the rest of this chapter. First, we take an evidence-first view. Where claims touch safety, we rely on recent peer-reviewed measurements or well-supported ablations rather than anecdote. The LiDAR spoofing and C-V2X denial-of-service results cited above are cases in point and they shape our discussion of defence-in-depth across sensing and communications [8, 9]. Second, we connect domains. Techniques proven in aerial or maritime autonomy often transfer to road vehicles with minimal adaptation, and a similar cross-pollination runs from mobile robotics to intelligent transport. The aquaculture AUV case study illustrates how autonomy works in cluttered, dynamic environments where visibility is poor and bandwidth is limited, constraints that also matter in urban driving [3] (Fig. 1).

The structure is as follows. We begin with sensing and perception for AIVs, then cover cooperation through V2X and CVIS. We next outline the threat model and

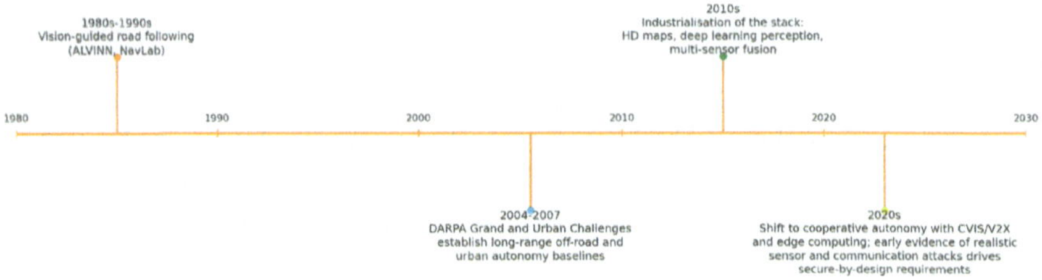

Fig. 1 Evolution timeline of AIVs

the state of defences across sensors, networks and middleware. We then examine decentralised trust and data provenance, including the suitability of permissioned blockchains for audit and update distribution in mobility systems. We close with open problems that draw together these strands, notably evaluation under coupled cyber-physical stressors, the privacy–utility balance in cooperative perception and governance for cross-operator interoperability. Key references throughout include foundational demonstrations of autonomy, recent surveys in wireless access and cooperative perception, security analyses for C-V2X and GNSS, and blockchain designs aimed at resource-constrained mobile agents.

2 Classification of AIVs

AIVs can be broadly classified according to their operational environment: ground-based, aerial-based, and water-based systems. This taxonomy provides a structured way to compare their design, autonomy requirements, and application domains. Ground vehicles such as self-driving cars and delivery robots operate within road and terrain constraints, aerial vehicles like drones and UAV swarms must navigate regulated airspace, while water-based systems, including autonomous ships and underwater vehicles, address maritime challenges. This classification, illustrated in Fig. 2,

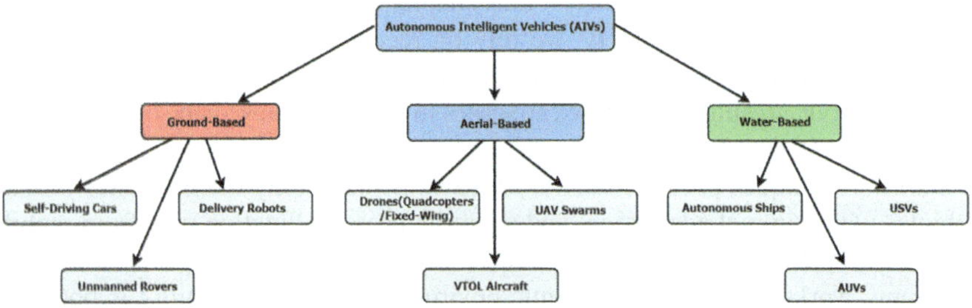

Fig. 2 Taxonomy of autonomous intelligent vehicles (AIVs)

establishes a foundation for analyzing their unique features, constraints, and use cases.

2.1 Aerial-Based AIVs

Unmanned Aerial Vehicles (UAVs) have become a central component of aerial-based AIVs due to their versatility, cost-effectiveness, and adaptability across diverse domains. Recent reviews highlight how advances in artificial intelligence have significantly extended UAV capabilities beyond traditional surveillance and mapping, enabling applications in traffic monitoring, disaster response, agriculture, and environmental conservation [19, 20]. The integration of machine learning and computer vision has facilitated autonomous navigation, real-time object detection, and predictive analytics, while generative AI models are increasingly being employed to forecast obstacles, optimize flight paths, and support adaptive mission planning. At the same time, green computing principles are being embedded in UAV design, combining energy-efficient algorithms and renewable power sources to extend endurance and reduce carbon emissions, thereby aligning UAV development with sustainability goals [21].

In addition to general-purpose roles, UAVs are making substantial inroads in specialized industrial applications such as power line inspection, where they provide a safer, more efficient, and cost-effective alternative to manual or helicopter-based surveys. Equipped with advanced sensors including LiDAR, thermal imaging, and optical cameras, UAVs can detect structural faults, monitor vegetation encroachment, and assess environmental risks in real time. Recent studies emphasize the role of AI-driven inspection algorithms, such as optimized YOLO variants and federated learning models, which improve fault detection accuracy while ensuring data privacy. These advances are complemented by innovations in robotic vision and autonomous control strategies that enhance UAV stability and navigation in complex or electromagnetically challenging environments [22]. Collectively, these developments position aerial-based AIVs as indispensable assets for infrastructure monitoring, urban management, and critical service delivery in both civilian and industrial contexts.

2.2 Ground-Based AIVs

Ground-based AIVs, encompassing self-driving cars, delivery robots, and unmanned rovers, are advancing rapidly due to progress in perception, navigation, and sensor fusion technologies. A central challenge in this domain is terrain traversability, particularly in unstructured or off-road environments where uncertainty in slope, surface friction, and obstacle density complicates autonomous navigation. Recent surveys emphasize the role of terrain classification, mapping, and cost-based traversability methods, which combine appearance-based, geometry-based, and hybrid approaches to improve decision-making under uncertain conditions [23]. Machine learning and deep learning algorithms now enable robust semantic segmentation of terrain types, while reinforcement learning supports adaptive decision-making across dynamic environments. Planetary rover missions, such as NASA's Curiosity and Perseverance, further demonstrate the necessity of reliable traversability analysis for safe navigation, reinforcing the importance of multi-sensor fusion that integrates LiDAR, cameras, and inertial measurement units.

In urban and commercial contexts, autonomous ground vehicles are increasingly deployed for last-mile logistics, where efficiency, cost-effectiveness, and sustainability are driving adoption. Studies comparing autonomous delivery vehicles with conventional and electric vans reveal that while upfront costs for electrified AGVs remain significant, the elimination of labor expenses and the integration of renewable energy technologies (e.g., solar-assisted charging) provide long-term economic advantages [24]. For example, autonomous parcel lockers mounted on AGVs offer scalable solutions to urban delivery challenges, reducing traffic congestion, minimizing environmental impacts, and improving delivery reliability. However, adoption also depends on infrastructural readiness, regulatory acceptance, and consumer trust in fully driverless operations. Collectively, these findings suggest that ground-based AIVs are not only pushing the frontier of autonomous mobility in extreme terrains but are also emerging as a cornerstone of sustainable urban logistics and industrial automation.

2.3 *Water-Based AIVs*

Water-based AIVs encompass a wide spectrum of platforms, including autonomous surface vehicles (ASVs), unmanned surface vessels (USVs), remotely operated vehicles (ROVs), and autonomous underwater vehicles (AUVs). These systems are increasingly deployed for applications such as oceanographic research, infrastructure inspection, maritime security, and environmental monitoring. ROVs, typically tethered to a surface vessel, excel in tasks requiring precise human-in-the-loop manipulation at significant depths, such as subsea pipeline inspection or offshore engineering. AUVs, by contrast, operate untethered and rely on preprogrammed missions and onboard autonomy to survey large ocean regions, perform seabed mapping, or conduct defense-related reconnaissance. Classification schemes often distinguish vehicles by propulsion method and autonomy: underwater gliders achieve extraordinary endurance by adjusting buoyancy rather than relying on propellers, while propelled AUVs combine onboard computation with thruster-based mobility for deep-sea missions. Recent surveys emphasize that design challenges for both ROVs and AUVs include power endurance, hull pressure resistance, underwater communication limitations, and the integration of AI-based perception systems for adaptive control [25, 26].

Alongside subsea platforms, USVs and hybrid vehicles have become crucial in bridging surface and underwater operations. USVs, operating on the water's surface,

Table 1 Comparison of ground-based, aerial-based, and water-based AIV domains with representative platforms, autonomy levels, communication requirements, and domain-specific security challenges

Domain	Representative platforms	Autonomy levels (SAE/equivalent)	Communication needs	Key security challenges
Ground-based	Self-driving cars, delivery robots, rovers	SAE Levels 0–5 (road vehicles); semi-autonomous for rovers	V2V, V2I, GPS, LiDAR, 5G/DSRC	GPS spoofing, sensor tampering, V2X data injection, privacy leakage
Aerial-based	Drones (UAVs), swarms, VTOL aircraft	Partial–full autonomy (under aviation regulations)	UAV-to-UAV, UAV-to-GCS, satellite, 5G	GPS jamming, hijacking of control links, swarm coordination attacks
Water-based	Autonomous ships, USVs, AUVs	Mission-specific autonomy (surface: semi to full; underwater: preprogrammed)	V2V maritime comms, satellite, acoustic/sonar links	Communication interception, navigation spoofing, maritime law compliance gaps

are increasingly used for hydrographic surveys, surveillance, and logistics, benefiting from extended range and easier integration with satellite communications. Recent work has also highlighted the role of multi-modal perception datasets for ASVs, combining LiDAR, radar, and vision sensors to improve situational awareness in unstructured maritime environments, thereby enhancing collision avoidance and autonomous navigation [27]. Hybrid vehicles such as ROV–AUV systems and UAV–AUV configurations further extend operational flexibility, enabling missions that shift between tethered and autonomous modes or even cross-domain operations between air and water. Bionic underwater robots, inspired by fish and cephalopods, add energy-efficient, stealthy alternatives for ecological monitoring and naval applications. Collectively, these innovations underscore the breadth of water-based AIVs and their critical role in advancing the vision of a "smart ocean," where sustainable resource management, defense, and climate monitoring increasingly depend on persistent, adaptive, and intelligent marine autonomy [26].

To summarize the distinguishing features across domains, Table 1 compares representative platforms, autonomy levels, communication needs, and domain-specific security challenges of ground, aerial, and water-based AIVs.

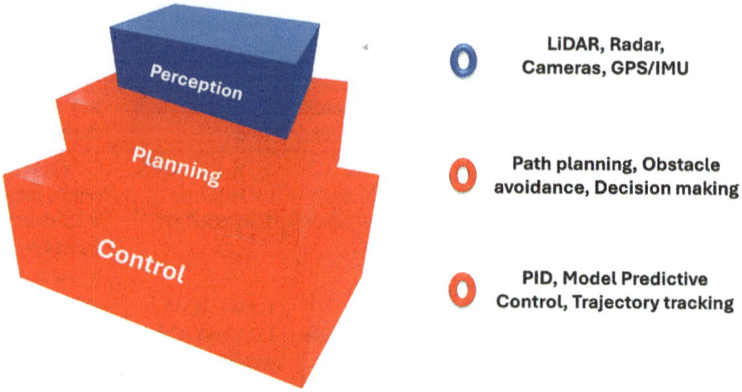

Fig. 3 Layered technology stack for AIVs

3 Core Technology Stack: Sensing, Planning and Control

Although ground, aerial and maritime AIVs vary in hardware, their architectures share a layered stack composed of *perception*, *planning* and *control* [28]. Figure 3 illustrates these layers. Each layer comprises subsystems that transform raw sensor data into actuation commands.

3.1 Perception: Sensors and Fusion

The perception layer converts multi-modal sensor streams into environment models. Cameras provide high-resolution images, but their performance degrades under low light or adverse weather [29, 30]. LiDAR sensors emit laser pulses and measure time-of-flight to build precise 3-D maps; however, they are expensive and can be spoofed, and cooperative perception research shows that attackers can "create erroneous obstacles or conceal real obstacles" through LiDAR spoofing, leading to unsafe behaviour if not detected [31, 32]. Radar offers long-range detection and is robust to weather but produces low-resolution point clouds [32]. Ultrasonic sensors support near-field detection for parking and docking. Many systems also use inertial measurement units (IMUs), global navigation satellite systems (GNSS) and barometers [30]. Sensor fusion algorithms (e.g., extended Kalman filters and deep learning models) combine these modalities to produce accurate localisation and object lists [33, 34].

Advanced perception relies on machine learning and high-performance computing [30, 35]. Convolutional neural networks (CNNs) and transformers perform object detection, semantic segmentation and scene understanding [30]. Edge computing hardware accelerates inference, while remote servers can augment computation when latency and bandwidth allow [35]. Emerging research explores semantic communication—transmitting only informative features instead of raw sensor data—and neuromorphic sensors that produce event-based outputs [29, 36, 37]. The Wevolver report emphasises that progress in AI algorithms, natural-language processing and machine learning is driving AIV capabilities [35].

3.2 Planning: Prediction, Behaviour and Path Generation

Planning translates perception outputs into trajectories and high-level decisions. Prediction modules estimate the future trajectories of nearby objects using kinematic models or deep recurrent networks [38]. Behaviour planners choose manoeuvres such as lane changes, merging or overtaking [39]. Path-planning algorithms search for collision-free routes; common methods include graph-based algorithms, sampling-based algorithms (rapidly exploring random trees), dynamic programming, meta-heuristic optimisation and nonlinear optimisation [40–42]. Many commercial systems still rely on heuristic rule sets, while research prototypes use learning-based approaches that optimise end-to-end reward functions [39, 43].

3.3 Control: Converting Plans Into Actions

Control modules convert desired paths into actuator commands. Classic proportional–integral–derivative (PID) controllers and adaptive controllers remain widely used in industry because they do not require precise models. Research communities, however, favour model-predictive control (MPC). Data-driven MPC "outperforms traditional PID and standard MPC" by learning system dynamics from data and reduces modeling effort [44]. Aerial vehicles often employ cascaded PID loops for attitude stabilisation and use linear quadratic regulators or energy-based control for trajectory tracking. Maritime vehicles must compensate for hydrodynamic forces, and controllers incorporate wave and current models. Increasingly, reinforcement learning algorithms are being tested for control tasks, but they face challenges related to safety guarantees and sample efficiency [39, 43].

3.4 Software and Hardware Infrastructure

Real-time operation requires robust computing platforms [35]. Automotive AIVs use heterogeneous system-on-chips with CPUs, GPUs and field-programmable gate arrays (FPGAs). Middleware such as the Robot Operating System (ROS) provides modularity and communication [35]. Deterministic execution is critical for meeting functional safety standards (e.g., ISO 26262). Many aerial and maritime vehicles run custom flight control firmware. Operating systems integrate cybersecurity features such as secure boot, memory protection and sandboxing [45]. As tasks become more computationally intensive, off-board cloud and edge resources support heavier workloads; however, this introduces new communication dependencies [35, 46].

4 Communication Networks in AIVs

Autonomous vehicles rarely operate in isolation. They exchange messages with other vehicles (V2V), infrastructure (V2I), pedestrians (V2P) and cloud back-ends (V2C). These interactions enable cooperative perception, platooning, dynamic traffic management and remote monitoring. Figure 4 illustrates a typical V2X architecture where vehicles communicate through dedicated short-range communications and cellular channels and where a cloud back-end aggregates data.

4.1 DSRC Versus C-V2X

Early V2V systems adopted dedicated short-range communications (DSRC) based on IEEE 802.11p; DSRC uses a carrier-sense multiple access (CSMA) protocol with asynchronous frames, leading to potential collisions in high-density scenarios [47, 48]. Cellular V2X (C-V2X) emerged to leverage 3GPP LTE and 5G technologies [47,

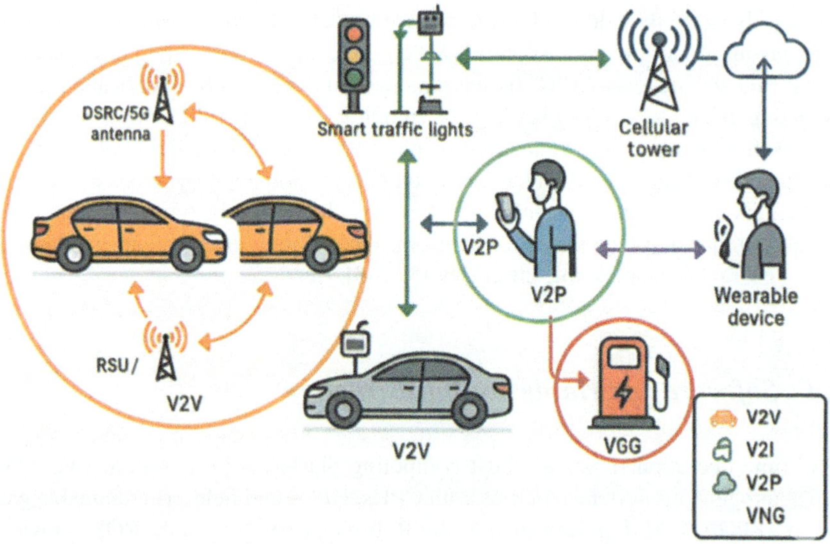

Fig. 4 V2X communication architecture

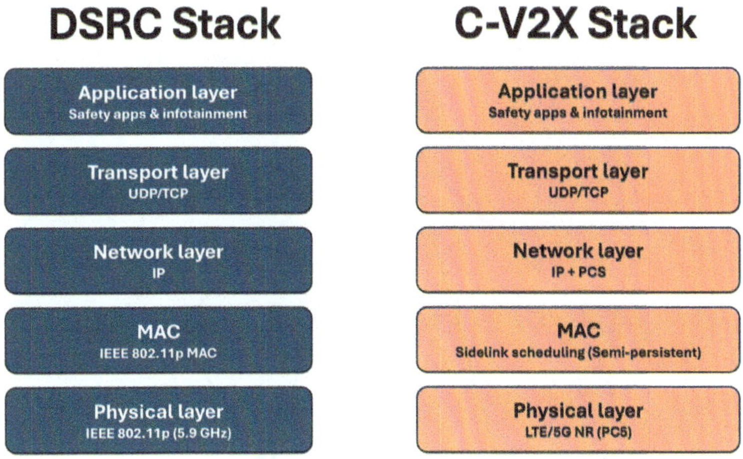

Fig. 5 Comparison of DSRC and C-V2X protocol stacks

49]. C-V2X divides the spectrum into resource blocks and uses tight time synchronisation via GNSS signals; this reduces collisions and improves range [47]. According to a 5G V2X survey, DSRC uses Wi-Fi-like CSMA while C-V2X uses resource-block scheduling, enabling better reliability and scalability [47]. C-V2X can operate in both direct (vehicle-to-vehicle) and network-based modes, whereas DSRC is primarily ad hoc [47, 49]. A comparative table shows that DSRC uses OFDM modulation and 10 MHz channels, whereas C-V2X employs LTE/5G modulation, supports 5 MHz and 10 MHz channels and benefits from higher transmit power [47]. These differences have led many governments to favour C-V2X and reallocate DSRC spectrum [47, 50].

Figure 5 contrasts the protocol stacks of DSRC and C-V2X. Both share a common application layer defined by SAE J2735 and support message types such as basic safety messages (BSMs), cooperative awareness messages and map data. However, their transport, network and physical layers differ, leading to distinct performance profiles.

4.2 5G/6G and Global Communication Frameworks

Modern AIVs increasingly rely on cellular networks for high-bandwidth, low-latency communication [46, 49]. Fifth-generation (5G) networks introduce ultra-reliable low-latency communications (URLLC), enhanced mobile broadband and network slicing; they support edge computing and multi-access edge computing (MEC) [46, 49]. A survey on 5G-enabled V2X emphasises that C-V2X and DSRC are being integrated into 5G, which provides better reliability through synchronous scheduling and supports high-data-rate sensor sharing [47, 49]. Sixth-generation (6G) networks

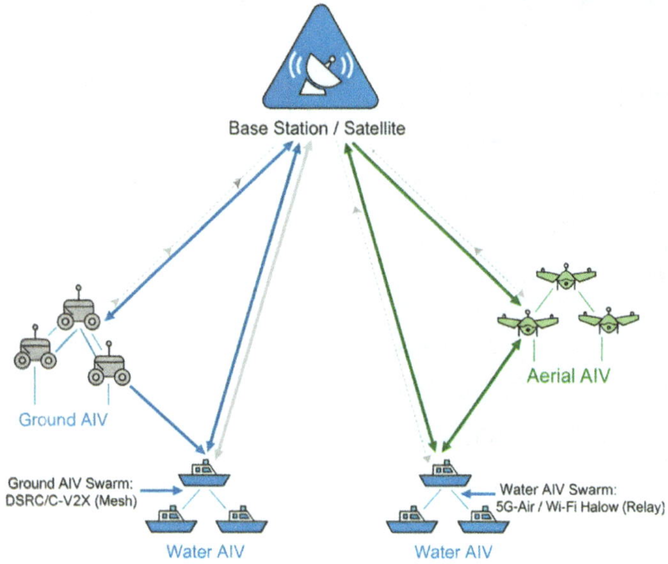

Fig. 6 Global communication framework for AIV swarms

promise terahertz bands, extreme reliability and integration of satellite and terrestrial networks [46, 49]. A 6G V2X survey notes that 6G will enable "ultra-low latency and high capacity" communications and use AI for dynamic resource management [49]. Combining terrestrial base stations, low-earth-orbit satellites and high-altitude platforms could provide global connectivity, enabling remote AIV operations in remote oceans or deserts [46, 49].

We propose a global communication framework for AIV swarms in Fig. 6. A base station or satellite network provides long-range connectivity; within each domain, vehicles exchange messages via local radio links. Inter-domain coordination (e.g., between ground vehicles and aerial drones) occurs via the back-end. Such frameworks are essential for cross-domain operations such as drone-assisted road surveillance or autonomous maritime logistics.

4.3 Physical-Layer Security and Jam Resistance

Dependence on wireless channels exposes AIVs to jamming, spoofing and eavesdropping [45, 51]. The 5G V2X survey acknowledges that new security models are needed because "physical-layer security based on massive multiple-input multiple-output (MIMO) antennas and beamforming" must be integrated with cryptographic protocols [45, 49]. Techniques such as spread spectrum, frequency hopping and cooperative diversity improve resilience [45]. Adaptive power control and beamforming can mitigate interference [45]. Emerging research explores machine-learning-based jamming detection and physical-layer key generation [9, 51].

5 Security Vulnerabilities in AIV Ecosystems

5.1 Framing

Autonomous Intelligent Vehicles (AIVs) operate as cyber-physical graphs. Vehicles, drones, vessels and roadside units exchange sensor features and control messages across imperfect channels while safety-critical decisions depend on machine learning components that can drift under distribution shift. We structure this section by three layers: Physical, Network, and Application or Machine Learning. The case studies and tables quantify impacts and show how weaknesses at one layer amplify risks in the others (Fig. 7).

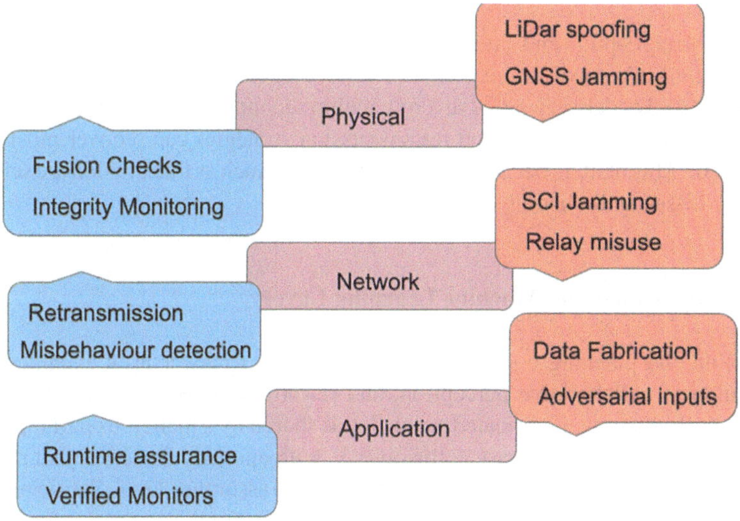

Fig. 7 Layered threat model and controls. The representative threats are in red and the primary controls in blue

5.2 *Threat Model by Layer*

5.2.1 Physical Layer

The most direct threats attempt to mislead or degrade sensing and localisation. Two families dominate current evidence. First is LiDAR spoofing and point removal or injection, which has now been demonstrated against vehicles in motion and with realistic hardware constraints [8, 52, 53]. Second is interference against satellite navigation such as GNSS jamming and spoofing, which propagates into degraded localisation and integrity if not monitored [11, 54–56].

Sensor fusion can mask isolated failures, yet the masking is not guaranteed when attacks are targeted at the dominant sensor for a given scene or when multiple sensors degrade together in poor weather or clutter [57]. The right question is not only whether a single sensor is robust, but whether the fusion graph can detect inconsistency quickly enough for safe braking.

5.2.2 Network Layer

Cooperative perception and safety beacons ride on cellular vehicle-to-everything (C-V2X) sidelink. Protocol-aware denial of service against sidelink control signalling can reduce packet delivery ratios severely within seconds under practical duty cycles, although Release 17 features and retransmission strategies can recover most of the loss [9, 10]. This matters because collaborative tasks such as track fusion are sensitive to bursty losses and schedule asymmetries.

5.2.3 Application and Machine Learning Layer

Perception and planning components are vulnerable to adversarial examples, data fabrication in collaborative perception, and reward or objective mis-specification. Recent work shows that fabricated point clouds shared by a malicious participant can induce hard braking or increase collision risk, with anomaly detection that recovers much of the degradation but not all of it [58]. At the assurance level, verification tied to threat models for cooperative messages and runtime monitoring can enforce safety and authenticity properties in use, rather than only at design time [59–61] (Fig. 8).

5.3 *Case Studies*

LiDAR spoofing against moving vehicles

Modern experiments have reached highway-relevant scenarios. Long-range tracking and high-frequency removal attacks achieve very high success rates up to braking distance at 60 km/h, and injection of ghost obstacles is feasible near the same distance.

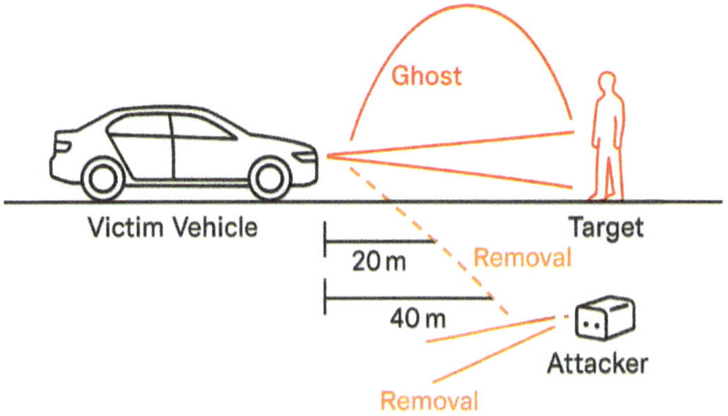

Fig. 8 LiDAR removal or injection geometry near braking distance: The attacker's device creates a false target ("Ghost") and attempts to remove a real object ("Target") from the vehicle's LiDAR, posing a serious safety risk

These results translate into stop-go failures and unnecessary hard braking if the vehicle does not validate sparse but safety-critical returns [52, 53]. Such results highlight the value of immutable sensor-data provenance: blockchain-based audit trails could ensure that fused perception data remain verifiable across vehicles, thereby reducing the impact of spoofed inputs. In a cooperative perception setting where roadside units (RSUs) and nearby AIVs share LiDAR frames, each node could publish hashed summaries of its sensor data to a lightweight blockchain ledger. When a spoofing or removal attack occurs, neighbouring vehicles can cross-check hashes to detect inconsistency in real time, enabling tamper-evident sensor sharing and post-incident forensics [62]. Table 2 summarises parameters that matter in practice and frames them as requirements for fusion and sanity checks (Fig. 9).

C-V2X sidelink denial of service and safety
Targeted jamming of sidelink control information can drive vehicle-to-vehicle delivery ratios down sharply in hardware-in-the-loop and simulation. When mapped into traffic-level simulations, the resulting increase in missed warnings and longer headway variance escalates safety risk, although Release 17 features such as aperiodic retransmission reduce the damage and raise effective delivery ratios again [9, 10]. A blockchain-backed reputation or scheduling mechanism could discourage persistent interference and distribute channel access decisions without relying on a single coordinator. In a secure cooperative platooning scenario, each vehicle could broadcast signed state updates such as speed, acceleration, and lane position to a shared ledger that orders messages through consensus. Even if sidelink communication experiences selective jamming, other platoon members can retrieve verified state information from the distributed ledger, maintaining safe inter-vehicle spacing and continuity of control. Table 3 links attack parameters to delivery and safety metrics and highlights control-level mitigations (Fig. 10).

Table 2 Physical and network evidence with operational implications

Attack type	Setup highlights	Observed impact	Implication
LiDAR point removal	Mobile victim at 60 km/h, long-range auto-aim, high-frequency removal [52]	$\geq$ 96% removal up to 20 m braking distance; injection can add $\geq 10^3$ points near 20 m	Require temporal sanity checks and cross-sensor veto when sparse pedestrian returns vanish or appear suddenly
LiDAR next-gen viability	New-generation LiDARs with pulse fingerprinting studied and attacked [53]	Feasible removal with grey-box timing and power limits	Expect partial diversity benefits but do not assume immunity
Sidelink control DoS	Targeted SCI jamming on C-V2X kits and USRPs [9]	PDR can fall by up to 90% for a target or channel PDR $<$ 25% within seconds; Release 17 aperiodic retransmission restores PDR to $\approx$ 85%	Provision aperiodic retransmission and misbehaviour detection in safety logic
Safety impact of DoS	Traffic-level mapping of DoS-induced losses [10]	Increased missed alerts and longer headway variance; quantified risk growth under load	Couple comms health to headway and braking policies

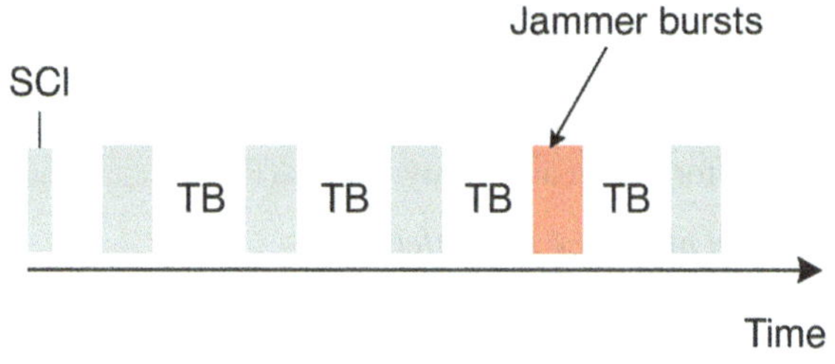

Fig. 9 A visual representation of how a jammer exploits the time-slotted nature of TDMA communication. While normal data packets (TB) are transmitted in their assigned slots, a targeted jammer burst corrupts a specific traffic burst, demonstrating a common form of signal interference

GNSS interference and integrity monitoring

Operational studies report an increase in wide-area jamming and spoofing episodes. Integrity monitoring and direction finding are practical first lines of defence, yet residual risk remains in urban canyons and during multi-path events. Integrating receiver autonomous integrity monitoring with inertial sensors and map constraints closes part of this gap [11, 54–56]. To further enhance trust, signed navigation attestations recorded on a shared ledger could enable vehicles to cross-verify localisation data and detect manipulation collectively. In maritime AIV fleets, for example, each vessel

Table 3 Mitigation levers and resource budgets

Layer	Mitigation	Mechanism	Budget and trade-off
Application	Temporal-spatial plausibility for cooperative perception [58]	Cross-agent anomaly detection on shared features or tracks	Low metadata overhead; small latency but large safety gain under attack
Network (C-V2X sidelink)	Release 17 aperiodic retransmission [9]	Decouple schedule predictability with one-shot resource selection	Modest extra airtime; measurable PDR recovery
Physical (localisation)	Integrity monitoring for GNSS [11, 56]	Receiver-level integrity checks, direction finding and INS aiding	Compute overhead on low-rate channels; protects localisation against silent drift
Application/runtime assurance	Runtime monitors tied to threat models [59, 61]	Monitors generated from verified properties and perception confidence	Small CPU budget if monitors focus on safety envelopes

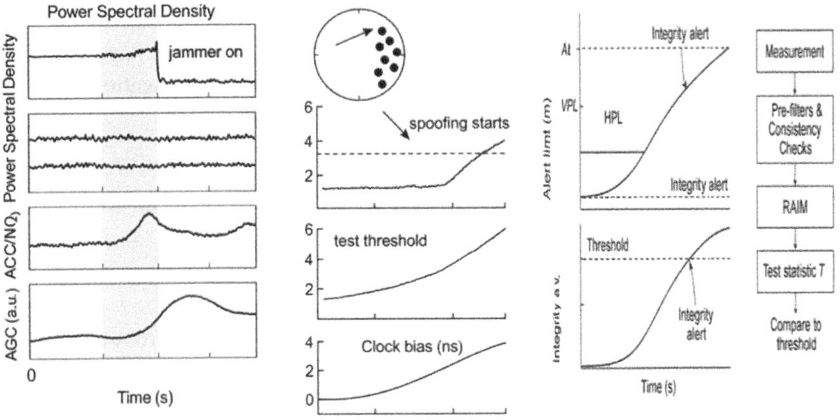

Fig. 10 The three panels show different methods a GNSS receiver uses to maintain integrity and detect attacks. From left to right: **Jamming Detection**: The left panel shows how a jammer, which broadcasts powerful noise, is detected. The Power Spectral Density and Automatic Gain Control (AGC) graphs show a clear increase when the jammer is on (gray shaded area) as the receiver's systems compensate for the interference. **Spoofing Detection**: The middle panel illustrates a spoofing attack, where a fake satellite signal misleads the receiver. The graph shows a detection metric crossing a threshold as the spoofer slowly "carries off" the receiver's internal clock, causing the Clock bias to drift. **Integrity Monitoring (RAIM)**: The right panel and flowchart explain Receiver Autonomous Integrity Monitoring (RAIM), a process that ensures the accuracy of the receiver's position solution. It shows how a growing signal error causes the calculated Protection Level to exceed an Alert Limit and a Test statistic to cross a Threshold, both of which trigger an integrity alert

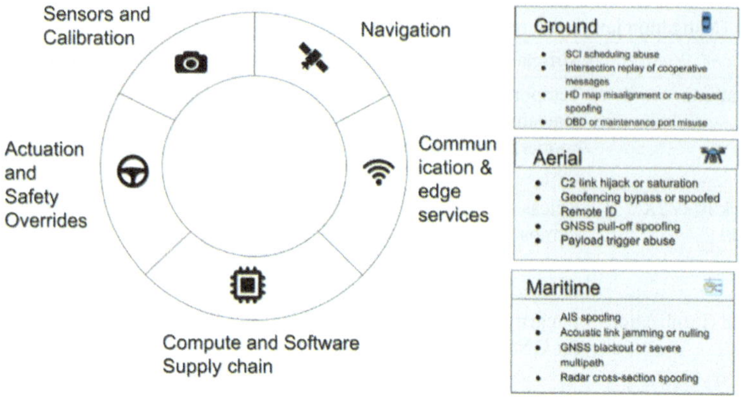

Fig. 11 Break-down the attack surface of a cyber-physical system, showing how vulnerabilities can be exploited within five core functional domains. The outer boxes provide concrete examples of attacks targeting different environments, from car-based GPS spoofing to drone hijacking and maritime radar deception

could periodically publish GNSS signal-quality metrics to a consortium blockchain. When one vessel reports abnormal drift, others can corroborate or flag discrepancies using the shared record, collectively mitigating spoofed navigation data and improving overall situational integrity (Fig. 11).

5.4 Collaborative-Mission Attack Surface

Cooperative perception
False data injection trades bandwidth for verifiability. Lightweight cross-agent plausibility checks, spatial-temporal consistency, and downweighting outlier tracks cut risk with modest overheads [58].

Platooning and swarms
Timing manipulation and consensus bias can propagate through convoys or mixed ground–air teams. Consensus that is robust to partial synchrony and membership churn limits the blast radius of a compromised node.

Human–system interaction and forensics
Remote operations and incident analysis need audit trails that are tamper-evident and privacy-preserving. Section 5.6 links these needs to verifiable logging.

Swarm UAVs
In UAV swarm missions, coordination commands are typically issued through shared wireless links, making them vulnerable to spoofing or replay attacks that could desynchronise flight paths or disrupt formation control. A blockchain-based consensus

layer can validate and timestamp each command or mission update across participating drones before execution. This approach is similar to SwarnRaft, which maintains swarm coherence in GNSS-denied conditions using blockchain-inspired consensus mechanisms [63]. In addition, dynamic formation control schemes leveraging blockchain and smart contracts have been proposed to enforce safe reconfiguration and leader election among drones [64]. This distributed validation ensures that only authorised and agreed-upon instructions propagate through the swarm, preventing rogue control messages and maintaining trust among heterogeneous aerial nodes. Beyond security, the immutable record of task assignments and sensor observations enables transparent post-mission auditing and accountability for cooperative aerial operations.

5.5 Why Conventional IT Security Is Insufficient

AIVs face tight deadlines for perception, planning and actuation. Heavyweight cryptography and centralised trust anchors degrade time budgets or create single points of failure. Standards now require cyber security engineering across the full lifecycle and explicit type approval processes, which steer designs toward runtime assurance, least-privilege communications and event traceability [65, 66].

5.6 From Gaps to Design Requirements

The tables and figures indicate three design requirements. First, treat sparse but high-salience objects as a special class in fusion and enforce temporal continuity, rather than trusting a single scan. Second, tie communication health to longitudinal control so that missing messages widen headways and reduce reliance on cooperative tracks. Third, pair standards-driven security engineering [65, 66] with runtime assurance for collaborative tasks, since pre-deployment proofs do not cover data trust during operation. These requirements collectively motivate a decentralized trust substrate, *blockchain*, for tamper-evident logging, identity management, and cross-domain auditability in AIV networks.

6 Implications and Need for Blockchain-Based Security

The layered vulnerabilities detailed in Sect. 5 reveal the limits of traditional, centralized security in AIV ecosystems. To meet the design requirements identified there; verifiable data integrity, decentralized identity, and auditability, this section explores how blockchain provides these guarantees through consensus-driven trust and immutable record-keeping.

Attacks such as identity spoofing, false data injection, and denial-of-service can disrupt mission-critical operations and compromise public safety. These challenges highlight the limitations of traditional centralized security models, which often introduce single points of failure and struggle to adapt to dynamic, cross-domain environments. Blockchain technology, with its decentralized architecture, consensus-driven trust establishment, and immutable record-keeping, has emerged as a compelling alternative. By distributing trust among multiple nodes and enabling transparent auditability of all interactions, blockchain provides the foundation for securing AIV ecosystems while ensuring accountability and resilience.

6.1 Functional Implications of Blockchain for AIVs

6.1.1 Data Integrity and Immutability

Blockchain's most immediate contribution to AIV ecosystems lies in ensuring that data exchanged among vehicles and infrastructure cannot be altered once recorded. For instance, navigation logs, sensor outputs, and V2X communication records can be immutably stored on distributed ledgers, making them resistant to tampering or unauthorized modification. This property is particularly valuable in scenarios requiring forensic accountability after accidents or cyber incidents, where verifiable audit trails are indispensable. Frameworks such as DEMU-NAV highlight how blockchain can serve as the backbone for secure autonomous navigation, leveraging smart contracts to govern interactions between vehicles, humans, and infrastructure in a transparent yet privacy-preserving manner [67].

6.1.2 Authentication and Identity Management

Another critical implication of blockchain adoption in AIVs is its ability to streamline and secure authentication across domains. Traditional PKI systems introduce high overhead for key management and are vulnerable to certificate revocation and single authority compromise. Recent work has demonstrated blockchain-based cross-domain authentication for UAV-assisted vehicular networks, where UAVs act as intermediaries while blockchain provides the trust anchor [68]. This approach reduces computational delays by at least 20% while maintaining privacy through biometric keys and Physical Unclonable Functions (PUFs), illustrating how decentralized identity frameworks can better support high-density and latency-sensitive mobility systems.

6.1.3 Decentralized Trust

The decentralized nature of blockchain eliminates reliance on central servers or ground control stations, which are attractive targets for adversaries. In UAV ad hoc networks (FANETs), blockchain has been used to distribute trust across multiple nodes, enabling secure coordination of drone swarms without a central controller [69]. Such decentralization is particularly valuable in contested or disaster-stricken environments, where communication infrastructure may be unreliable or absent. By embedding consensus mechanisms directly into the coordination layer, blockchain ensures that no single compromised node can destabilize the entire network.

6.1.4 Resilience and Availability

Blockchain also strengthens resilience by providing redundancy across distributed ledgers. Even if certain nodes fail or are compromised, the integrity of the ledger remains intact, ensuring continuous availability of critical security services such as authentication and data verification. Studies on blockchain-enhanced UAV communication demonstrate how ledger replication and lightweight consensus mechanisms can reduce downtime while maintaining system throughput [70]. This property is crucial for mission-critical AIV applications, from cooperative platooning on highways to coordinated UAV swarms in disaster zones.

6.1.5 Interoperability Across Domains

Finally, blockchain addresses interoperability challenges that arise when AIVs from different domains, ground, aerial, and maritime, must securely interact. For example, UAV-assisted vehicular networks integrate blockchain to harmonize authentication and data exchange across heterogeneous environments [68]. By serving as a common trust layer, blockchain facilitates consistent security policies and identity verification across domains, overcoming the fragmentation of traditional security models. This interoperability is foundational to realizing multi-domain autonomous transport systems where cars, drones, and ships collaboratively share information in real time.

6.2 The Need for Blockchain in AIV Security

Traditional centralized security models in vehicular and UAV networks introduce bottlenecks and single points of failure, making them ill-suited for the dynamic and high-density environments where AIVs operate. Blockchain addresses these weaknesses by decentralizing authentication and trust, enabling faster and more resilient operations. For example, blockchain-based cross-domain authentication in UAV-assisted vehicular systems has shown reductions in computational and communication delays of at least 20%, highlighting its necessity for real-time safety-critical applications [68].

Beyond resilience, blockchain provides the transparency and auditability required for trust in autonomous systems. By maintaining immutable records of navigation data, communications, and mission logs, blockchain creates verifiable trails that can be used for forensic analysis after accidents or disputes. Frameworks such as DEMU-NAV demonstrate how smart contracts can govern roles and permissions across vehicles, humans, and regulatory authorities, ensuring accountability while preserving privacy [67]. This capability not only secures technical operations but also strengthens public and regulatory trust in AIV adoption.

Finally, blockchain offers scalability and adaptability for the rapidly growing AIV ecosystem. Lightweight consensus protocols and decentralized trust distribution allow blockchain-based UAV and vehicular networks to function securely in high-density, real-time environments [69]. By enabling consistent authentication and coordination across ground, aerial, and maritime domains, blockchain becomes more than a supportive technology, it is an essential foundation for securing the next generation of intelligent, interconnected mobility systems [68].

6.3 Challenges and Caveats

While blockchain offers clear security benefits for AIV ecosystems, its adoption is not without limitations. Consensus mechanisms such as Proof of Work (PoW) and even some Proof of Stake (PoS) variants introduce latency and high computational costs that are incompatible with the real-time demands of autonomous navigation. In resource-constrained platforms like UAVs, where energy and bandwidth are scarce, these overheads can significantly degrade performance [70].

Another challenge lies in balancing scalability with efficiency. Although blockchain can support decentralized authentication and data sharing across domains, performance tends to deteriorate as transaction volumes increase. High-density vehicular or UAV swarm environments require lightweight protocols and optimized consensus schemes to avoid congestion and maintain low latency [68]. Furthermore, ensuring privacy in transparent ledgers remains an open issue, as public visibility of transactions can inadvertently expose sensitive trajectory or operational data if not carefully managed. Table 4 provides a layered perspective on AIV threats, their conventional mitigations, and the added contributions of blockchain, illustrating how decentralized trust mechanisms reinforce resilience across physical, network, and application layers.

Finally, integrating blockchain into heterogeneous AIV domains introduces interoperability and standardization challenges. Ground vehicles, aerial drones, and maritime platforms each operate under different regulatory and technical frameworks, complicating unified blockchain adoption. Solutions such as permissioned blockchains and domain-specific smart contracts have been proposed, but these

Table 4 Layered threats and mitigation strategies in AIV ecosystems

Layer	Typical threats	Mitigation strategies	Blockchain contributions
Physical layer	GPS spoofing, sensor jamming, hardware tampering	Redundant sensing, hardened GNSS, secure hardware modules	Immutable sensor logs, trusted timestamping
Network layer	DoS/DDoS attacks, Sybil attacks, eavesdropping	Secure routing, intrusion detection, encrypted comms	Decentralized trust, consensus-based validation of nodes
Application layer	Data tampering, privacy breaches, unauthorized access	Access control, anomaly detection, AI-based intrusion detection	Smart contracts for authentication, privacy-preserving data sharing

must be carefully designed to remain interoperable and scalable [67]. Addressing these caveats is critical if blockchain is to serve as a reliable security backbone for autonomous intelligent vehicles.

7 Conclusion and Future Directions

Autonomous Intelligent Vehicles (AIVs) epitomize the convergence of robotics, artificial intelligence, communications, and cybersecurity. This chapter has provided a holistic perspective on their evolution from sensing and perception to cooperative communication networks, threat landscapes, and decentralized trust mechanisms. By classifying AIVs across ground, aerial, and maritime domains and examining their technological and security architectures, we have shown how these systems are transitioning from isolated agents into tightly networked, cyber-physical ecosystems. The analysis highlights that modern AIV ecosystems demand resilience not only against traditional cyber threats but also against safety-critical vulnerabilities at the physical and decision layers. As vehicles collaborate via C-V2X and edge networks, security assurance must extend across sensors, communication links, and machine-learning pipelines. Blockchain emerges as a unifying trust fabric that complements standards-driven cybersecurity by ensuring verifiable data integrity, transparent auditing, and decentralized control.

Blockchain strengthens AIV assurance through four key contributions: decentralizing identity and access using DIDs/VCs to eliminate single points of failure in C–V2X, maintaining tamper-evident, auditable logs of perception, control, and maintenance data for forensics and regulatory compliance, enabling policy-based on-chain governance for rapid credential revocation and misbehavior reporting during attacks, anchoring provenance for models, updates, and cooperative perception artifacts to limit the impact of falsified data. Implemented through permissioned ledgers and edge checkpoints, these mechanisms enhance trust and accountability across cooperative platoons, UAV swarms, and maritime logistics without exceeding real-time latency budgets. In essence, the secure evolution of AIVs depends on merging traditional control theory and communication reliability with decentralized, cryptographic, and AI-driven assurance mechanisms. Together, they form the foundation for scalable, auditable, and trustworthy autonomy that can safely transform transportation, logistics, and connected infrastructure.

The next phase of AIV research will emphasize deeper integration of communications, computing, and security. Sixth-generation (6G) networks promise terabit-per-second data rates, sub-millisecond latency, and unified terrestrial-satellite coverage to support global autonomy. Digital twins will enable large-scale simulation and optimization before deployment, while semantic and intent-based communications will minimize bandwidth demands. Equally crucial, *quantum-safe cryptography* and *quantum key distribution* will secure AIV systems against future quantum adversaries. From a policy standpoint, harmonized standards for V2X protocols, cybersecurity certification, and data privacy must evolve alongside technology. Public engagement and interdisciplinary education will remain vital for ensuring transparency, accountability, and long-term trust in autonomous mobility systems.

References

1. Thrun, S., Montemerlo, M., Dahlkamp, H., Stavens, D., Aron, A., Diebel, J., Fong, P., Gale, J., Halpenny, M., Hoffmann, G., Lau, K., Oakley, C., Palatucci, M., Pratt, V., Stang, P., Strohband, S., Dupont, C., Jendrossek, L.-E., Koelen, C., Markey, C., Rummel, C., Niekerk, J., Jensen, E., Alessandrini, P., Bradski, G., Davies, B., Ettinger, S., Kaehler, A., Nefian, A., Mahoney, P.: Stanley: the robot that won the darpa grand challenge. J. Field Robot. **23**(9), 661–692 (2006). https://doi.org/10.1002/rob.20147
2. Urmson, C., Anhalt, J., Bagnell, D., Baker, C., Bittner, R., Clark, M.N., Dolan, J., Duggins, D., Galatali, T., Geyer, C., Gittleman, M., Harbaugh, S., Hebert, M., Howard, T.M., Kolski, S., Kelly, A., Likhachev, M., McNaughton, M., Miller, N., Peterson, K., Pilnick, B., Rajkumar, R., Rybski, P., Salesky, B., Seo, Y.-W., Singh, S., Snider, J., Stentz, A., Whittaker, W., Wolkowicki, Z., Ziglar, J., Bae, H., Brown, T., Demitrish, D., Litkouhi, B., Nickolaou, J., Sadekar, V., Zhang, W., Struble, J., Taylor, M., Darms, M., Ferguson, D.: Autonomous driving in urban environments: boss and the urban challenge. J. Field Robot. **25**(8), 425–466 (2008). https://doi.org/10.1002/rob.20255
3. Peck, C.J., Langedock, K., Boone, W., Fourie, F., Moulaert, I., Semeraro, A., Sterckx, T., Geldhof, R., Groenendaal, B., Ponsoni, L.: The use of autonomous underwater vehicles for monitoring aquaculture setups in a high-energy shallow water environment: case study Belgian North Sea. Front. Mar. Sci. **11**, (2024). https://doi.org/10.3389/fmars.2024.1386267
4. Ajakwe, S.O., Ihekoronye, V.U., Kim, D.-S., Lee, J.-M.: Alien: assisted learning invasive encroachment neutralization for secured drone transportation system. Sensors **23**(3), 1233 (2023)
5. Ji, Y., Zhou, Z., Yang, Z., Huang, Y., Zhang, Y., Zhang, W., Xiong, L., Yu, Z.: Toward autonomous vehicles: a survey on cooperative vehicle-infrastructure system. Iscience **27**(5), 109751 (2024). https://doi.org/10.1016/j.isci.2024.109751
6. Zimmer, W., Wardana, G.A., Sritharan, S., Zhou, X., Song, R., Knoll, A.C.: Tumtraf v2x cooperative perception dataset. In: 2024 IEEE/CVF Conference on Computer Vision and Pattern

Recognition (CVPR), pp. 22668–22677 (2024). https://doi.org/10.1109/CVPR52733.2024.02139
7. Ajakwe, S.O., Igboanusi, I.S., Lee, J.-M., Kim, D.-S.: iBANDA: a blockchain-assisted defense system for authentication in drone-based logistics. Drones **9**(8), 590 (2025)
8. Suzuki, R., Sato, T., Hayakawa, Y., Ikeda, K., Sako, O., Nagata, R., Chen, Q.A., Yoshioka, K.: Wip: Towards practical lidar spoofing attack against vehicles driving at cruising speeds. In: Symposium on Vehicles Security and Privacy (VehicleSec) 2024. NDSS Symposium (2024). https://doi.org/10.14722/vehiclesec.2024.23023
9. Twardokus, G., Rahbari, H.: Toward protecting 5g sidelink scheduling in c-v2x against intelligent dos attacks. IEEE Trans. Wirel. Commun. **22**(11), 7273–7286 (2023). https://doi.org/10.1109/TWC.2023.3249665
10. Pethő, Z., Kazár, T.M., Szalay, Z., Török, A.: Quantifying cyber risks: the impact of dos attacks on vehicle safety in v2x networks. IEEE Trans. Intell. Transp. Syst. **25**(11), 18591–18600 (2024). https://doi.org/10.1109/TITS.2024.3436840
11. Qiao, J., Lu, Z., Lin, B., Song, J., Xiao, Z., Wang, Z., Li, B.: A survey of GNSS interference monitoring technologies. Front. Phys. **11**, 1133316 (2023). https://doi.org/10.3389/fphy.2023.1133316
12. Clancy, J., Mullins, D., Deegan, B., Horgan, J., Ward, E., Eising, C., Denny, P., Jones, E., Glavin, M.: Wireless access for v2x communications: research, challenges and opportunities. IEEE Commun. Surv. & Tutor. **26**(3), 2082–2119 (2024). https://doi.org/10.1109/COMST.2024.3384132
13. Huang, T., Liu, J., Zhou, X., Nguyen, D.C., Azghadi, M.R., Xia, Y., Han, Q.-L., Sun, S.: Vehicle-to-everything cooperative perception for autonomous driving. Proc. IEEE 1–35 (2025). https://doi.org/10.1109/JPROC.2025.3600903
14. Liu, Y., Gao, J., Lu, Y., Cao, R., Yao, L., Xia, Y., Han, D.: Lightweight blockchain-enabled secure data sharing in dynamic and resource-limited UAV networks. IEEE Netw. Mag. Glob. Internetworking **38**(4), 25–31 (2024). https://doi.org/10.1109/MNET.2024.3383237
15. Rao, P.M., Jangirala, S., Pedada, S., Das, A.K., Park, Y.: Blockchain integration for IoT-enabled v2x communications: a comprehensive survey, security issues and challenges. IEEE Access **11**, 54476–54494 (2023). https://doi.org/10.1109/ACCESS.2023.3281844
16. BelMannoubi, S., Touati, H., Hadded, M., Toumi, K., Shagdar, O., Kamoun, F.: A comprehensive survey on blockchain-based c-its applications: Classification, challenges, and open issues. Veh. Commun. **43**, 100607 (2023). https://doi.org/10.1016/j.vehcom.2023.100607
17. Liu, Q., Xue, Y., Zhang, Y., Chen, D., Han, K.: AdaMap: high-scalable real-time cooperative perception at the edge. In: Proceedings of the Eighth ACM/IEEE Symposium on Edge Computing, pp. 14–26. Association for Computing Machinery (2024). https://doi.org/10.1145/3583740.3626612
18. Hawlader, F., Robinet, F., Frank, R.: Leveraging the edge and cloud for v2x-based real-time object detection in autonomous driving. Comput. Commun. **213**(C), 372–381 (2024). https://doi.org/10.1016/j.comcom.2023.11.025
19. Ajakwe, S.O., Kim, D.-S.: Facets of security and safety problems and paradigms for smart aerial mobility and intelligent logistics. IET Intell. Transp. Syst. **18**, 2827–2855 (2024)
20. Ajakwe, S.O., Kim, D.-S., Lee, J.-M.: Drone transportation system: systematic review of security dynamics for smart mobility. IEEE Internet Things J. **10**(16), 14462–14482 (2023)
21. Pal, O.K., Shovon, M.S.H., Mridha, M., Shin, J.: In-depth review of ai-enabled unmanned aerial vehicles: trends, vision, and challenges. Discov. Artif. Intell. **4**(1), 97 (2024)
22. Mendu, B., Mbuli, N.: State-of-the-art review on the application of unmanned aerial vehicles (UAVs) in power line inspections: Current innovations, trends, and future prospects. Drones **9**(4), 265 (2025)
23. Beycimen, S., Ignatyev, D., Zolotas, A.: A comprehensive survey of unmanned ground vehicle terrain traversability for unstructured environments and sensor technology insights. Eng. Sci. Technol. Int. J. **47**, 101457 (2023)
24. Levkovych, O., Saraceni, A.: Efficiency in the last mile of autonomous ground vehicles with lockers: from conventional to renewable energy transport. Sustainability **15**(23), 16219 (2023)

25. Ghatak, A., Mukherjee, K., Singh, Y.: Classification on unmanned underwater vehicles: a review. Int. J. Veh. Auton. Syst. **17**(3–4), 192–220 (2023)
26. Xu, J., Irigoien, X., Alouini, M.-S.: State-of-the-art underwater vehicles and technologies enabling smart ocean: survey and classifications (2024). arXiv:2412.18667
27. Jeong, M., Chadda, A., Ren, Z., Zhao, L., Liu, H., Zhang, A., Jiang, Y., Achong, S., Lensgraf, S., Roznere, M., et al.: Seepersea: multi-modal perception dataset of in-water objects for autonomous surface vehicles. IEEE Trans. Field Robot. (2025)
28. Naz, N., Ehsan, M.K., Amirzada, M.R., Ali, M.Y., Qureshi, M.A.: Intelligence of autonomous vehicles: a concise revisit. J. Sens. **2022**(1), 2690164 (2022). https://doi.org/10.1155/2022/2690164
29. Zhou, B., Jiang, J.: Deep event-based object detection in autonomous driving: a survey (2024). arXiv:2405.03995
30. Pickeral, A., Marquez, E., Smith, M., Calhoun, J.: Efficient vision transformers for autonomous off-road perception systems. J. Comput. Commun. **12**, 188–207 (2024). https://doi.org/10.4236/jcc.2024.129011
31. Zhang, H., Li, Z., Cheng, S., Clark, A.: Cooperative perception for safe control of autonomous vehicles under LiDAR spoofing attacks (2023). arXiv:2302.07341
32. Ersü, C., Petlenkov, E., Janson, K.: A systematic review of cutting-edge radar technologies: applications for unmanned ground vehicles (UGVs). Sensors (Basel) **24**(23), 7807 (2024). https://doi.org/10.3390/s24237807
33. Yu, Z., Wan, W., Ren, M., Zheng, X., Fang, Z.: Sparsefusion3d: sparse sensor fusion for 3d object detection by radar and camera in environmental perception. IEEE Trans. Intell. Veh. **9**(1), 1524–1536 (2024). https://doi.org/10.1109/TIV.2023.3331972
34. Kwak, J., Jeon, H., Song, S.: Efficient multi-sensor fusion for cooperative autonomous vehicles leveraging c-its infrastructure and machine learning. Sensors **25**(7), (2025). https://doi.org/10.3390/s25071975
35. Islayem, R., Alhosani, F., Hashem, R., Alzaabi, A., Meribout, M.: Hardware accelerators for autonomous cars: a review (2024). arXiv:2405.00062
36. Ye, S., Wu, Q., Fan, P., Fan, Q.: A survey on semantic communications in internet of vehicles (2025). arXiv:2503.03767
37. Meng, R., Gao, S., Fan, D., Gao, H., Wang, Y., Xu, X., Wang, B., Lv, S., Zhang, Z., Sun, M., Han, S., Dong, C., Tao, X., Zhang, P.: A survey of secure semantic communications (2025). arXiv:2501.00842
38. Xu, M., Liu, Z., Wang, B., Li, S.: A survey of autonomous driving trajectory prediction: methodologies, challenges, and future prospects. Machines **13**(9), (2025). https://doi.org/10.3390/machines13090818
39. Klimke, M., Völz, B., Buchholz, M.: Integration of reinforcement learning based behavior planning with sampling based motion planning for automated driving. In: 2023 IEEE Intelligent Vehicles Symposium (IV), pp. 1–8. IEEE (2023). https://doi.org/10.1109/iv55152.2023.10186736
40. Ji, Y., Sun, R., Wang, Y., Zhu, Z., Liu, Z.: Dynamic path planning for unmanned autonomous vehicles based on CAS-UNet and graph neural networks. Sensors (Basel) **25**(14), 4283 (2025). https://doi.org/10.3390/s25144283
41. Ugwoke, K.C., Nnanna, N.A., Abdullahi, S.E.-Y.: Simulation-based review of classical, heuristic, and metaheuristic path planning algorithms. Sci. Rep. **15**(1), 12643 (2025). https://doi.org/10.1038/s41598-025-96614-2
42. Rathi, A., Geetha, R., Vadali, M.: Dynamic prioritization for conflict-free path planning of multi-robot systems (2021). arXiv:2101.01978
43. Li, Z., Jin, G., Yu, R., Chen, Z., Li, N., Han, W., Xiong, L., Leng, B., Hu, J., Kolmanovsky, I., Filev, D.: A survey of reinforcement learning-based motion planning for autonomous driving: lessons learned from a driving task perspective (2025). arXiv:2503.23650
44. Giannini, F., Famularo, D.: A data-driven approach to set-theoretic model predictive control for nonlinear systems. Information **15**(7), (2024). https://doi.org/10.3390/info15070369

45. He, C., Wang, W., Jiang, W., He, Z., Wang, J., Xie, X.: Security for the internet of vehicles with integration of sensing, communication, computing, and intelligence: a comprehensive survey. Sensors **25**, 5119 (2025). https://doi.org/10.3390/s25165119
46. Chataut, R., Nankya, M., Akl, R.: 6g networks and the ai revolution—exploring technologies, applications, and emerging challenges. Sensors (Basel) **24**(6), 1888 (2024). https://doi.org/10.3390/s24061888
47. Wang, J., Topilin, I., Feofilova, A., Shao, M., Wang, Y.: Cooperative intelligent transport systems: the impact of C-V2X communication technologies on road safety and traffic efficiency. Sensors **25**(7), 2132. https://doi.org/10.3390/s25072132
48. Okafor, C.I., Ahakonye, L.A.C., Lee, J.M., Kim, D.-S.: Purequantum: towards a scalable blockchain channel security in iot networks. Blockchain Res. Appl. 100372 (2025). https://doi.org/10.1016/j.bcra.2025.100372
49. Wang, D., Qiu, A., Zhou, Q., Schotten, H.D.: A survey on the role of artificial intelligence and machine learning in 6G-V2X applications (2025). arXiv:2506.09512
50. Jin, Y., Liu, X., Zhu, Q.: DSRC & C-V2X comparison for connected and automated vehicles in different traffic scenarios (2022). arXiv:2203.12553
51. Panitsas, I., Yigit, Y., Tassiulas, L., Maglaras, L., Canberk, B.: JamShield: a machine learning detection system for over-the-air jamming attacks (2025). arXiv:2507.11483
52. Sato, T., Suzuki, R., Hayakawa, Y., Ikeda, K., Sako, O., Nagata, R., Yoshida, R., Chen, Q.A., Yoshioka, K.: On the realism of LiDAR spoofing attacks against autonomous driving vehicle at high speed and long distance. In: Network and Distributed System Security Symposium (NDSS) (2025)
53. Sato, T., Hayakawa, Y., Suzuki, R., Shiiki, Y., Yoshioka, K., Chen, Q.A.: LiDAR spoofing meets the new-gen: capability improvements, broken assumptions, and new attack strategies. In: Network and Distributed System Security Symposium (NDSS) (2024). https://doi.org/10.14722/ndss.2024.23350
54. Lu, C., Lu, Z., Liu, Z., Huang, L., Chen, F.: Overview of satellite nav spoofing and anti-spoofing techniques. Front. Phys. **12**, 1428544 (2024). https://doi.org/10.3389/fphy.2024.1428544
55. Radoš, K., Brkić, M., Begušć, D.: Recent advances on jamming and spoofing detection in GNSS. Sensors **24**(13), 4210 (2024). https://doi.org/10.3390/s24134210
56. Huang, Z., Mou, W., Wang, R., Li, P., Lyu, Z., Ou, G.: A survey of GNSS receiver autonomous integrity monitoring: research status and opportunities. Front. Phys. **13**, 1567301 (2025). https://doi.org/10.3389/fphy.2025.1567301
57. Matos, F., Bernardino, J., Durães, J., Cunha, J.: A survey on sensor failures in autonomous vehicles: challenges and solutions. Sensors (2024)
58. Zhang, Q., Jin, S., Zhu, R., Sun, J., Zhang, X., Chen, Q.A., Mao, Z.M.: On data fabrication in collaborative vehicular perception: attacks and countermeasures. In: USENIX Security Symposium (2024)
59. Farrell, M., Bradbury, M., Cardoso, R.C., Fisher, M., Dennis, L.A., Dixon, C., Sheik, A.T., Yuan, H., Maple, C.: Security-minded verification of cooperative awareness messages. IEEE Trans. Dependable Secur. Comput. **21**(4), 4048–4065 (2024). https://doi.org/10.1109/TDSC.2023.3345543
60. Miller, K., Zeitler, C.K., Shen, W., Hobbs, K., Schierman, J., Viswanathan, M., Mitra, S.: Optimal runtime assurance via reinforcement learning. In: 2024 ACM/IEEE 15th International Conference on Cyber-Physical Systems (ICCPS), pp. 67–76 (2024). https://doi.org/10.1109/ICCPS61052.2024.00013
61. Pasquale Antonante, H.G.N., Carlone, L.: Monitoring of perception systems: deterministic, probabilistic, and learning-based fault detection and identification. Artif. Intell. (2023)
62. Andert, E., Mendoza, F., Behrens, H.W., Shrivastava, A.: Conclave—secure and robust cooperative perception for cavs using authenticated consensus and trust scoring. In: Proceedings of the 61st ACM/IEEE Design Automation Conference. DAC '24 (2024). https://doi.org/10.1145/3649329.3658491
63. Dev, K., Madhwal, Y., Shevelo, S., Osinenko, P., Yanovich, Y.: Swarnraft: leveraging consensus for robust drone swarm coordination in GNSS-degraded environments (2025). arXiv:2508.00622

64. Li, H., Li, P., Zhang, W., Shi, H., Tan, L., Zhang, P.: Secure communication and dynamic formation control of intelligent drone swarms using blockchain technology. Information **16**(9), 768 (2025). https://doi.org/10.3390/info16090768
65. ISO/SAE 21434: Road vehicles—Cybersecurity engineering. International Standard (2021)
66. UNECE Regulation No. 155: Cyber Security and Cyber Security Management System. Working documents and guidance (2025)
67. Anda-Suárez, J., López-Ramírez, J.L., Jimenez-Mendoza, D., Benitez-Quintero, J.M., Avina-Bravo, E.G., Gutierrez-Hernandez, D.A., Avina-Cervantes, J.G.: An integrated blockchain framework for secure autonomous vehicle communication system. Information **16**(7), 557 (2025)
68. Wang, W., Zhang, S., Liu, G., Zhao, Y.: A blockchain-based cross-domain authentication scheme for unmanned aerial vehicle-assisted vehicular networks. World Electr. Veh. J. **16**(4), 199 (2025)
69. Alqarni, M.A.: Secure UAV adhoc network with blockchain technology. PLoS ONE **19**(5), 0302513 (2024)
70. Hossain, M.I., Tahtali, M., Turhan, U., Biswas, K.: Blockchain integration in UAV networks: performance metrics and analysis. Sensors **24**(23), 7813 (2024)

Introduction to Blockchain and Autonomous Intelligent Vehicles

Douglas Allswell Kelechi, Lucky Uchenna Chibuike, Ukabuiro Ikenna Kelechi, and Chidi Ukamaka Betrand

Abstract The increasing use of autonomous intelligent vehicles (AIVs), which provide more efficiency, safety, and convenience, has led to tremendous advancements in the automobile sector. Additionally, blockchain technology has emerged as a secure, decentralised substitute for a variety of applications. However, the combination of AIVs with blockchain technology is still in its infancy; several problems, including those related to scalability, security, and data management, still need to be fixed. Effective authorisation, authentication, and data exchange systems are essential for the safe and reliable operation of Autonomous Intelligent Vehicles (AIVs). This chapter introduces the fundamental concepts of blockchain technology and discusses some potential applications in AIVs. We looked into the possible uses of blockchain-based solutions for safe data sharing and communication in AIVs. Additionally discussed were the challenges and limitations of integrating blockchain technology with AIVs. The potential benefits of integrating blockchain technology with AIVs, such as enhanced security, credibility, and data integrity, were illustrated by our study. It also highlighted important concerns that were handled, including scalability, latency, and regulatory frameworks. The marriage of blockchain technology with AIVs was thoroughly introduced in this chapter, with an emphasis on the potential benefits and challenges. To overcome the challenges and fully utilise

D. A. Kelechi (✉) · C. U. Betrand
Department of Computer Science, Federal University of Technology, Owerri, Nigeria
e-mail: kelechi.douglas@futo.edu.ng

C. U. Betrand
e-mail: chidi.betrand@futo.edu.ng

L. U. Chibuike
Directorate of General Studies, University of Agriculture and Environmental Sciences, Umuagwo, Nigeria
e-mail: uchenna.lucky@uaes.edu.ng

U. I. Kelechi
Department of Computer Science, Abia State University, Abia State, Uturu, Nigeria
e-mail: Ikenna.ukabuiro@abiastateuniversity.edu.ng

S. O. Ajakwe and D.-S. Kim (eds.), *Blockchain for Security of Autonomous Intelligent Vehicles*, Studies in Big Data 194,
https://doi.org/10.1007/978-3-032-20851-4_2

blockchain-based AIVs, the results showed how much more research and development are needed. Transportation, security, and data management were all significantly impacted by the combination of blockchain and AIVs.

Keywords Blockchain · Autonomous vehicles · Vehicle-to-everything · Security · Autonomous intelligent vehicles

1 Introduction

Vehicle communication systems must be dependable and secure as autonomous intelligent vehicles (AIVs) become a necessary component of contemporary transportation. In order to interact with urban infrastructure, autonomous intelligent cars must use networked systems, which present serious threats like impersonation, data manipulation, and cyber-attacks. These flaws may compromise AVs' security and ability to make decisions in real-time traffic scenarios.

Blockchain technology is one possible remedy. Chen et al. [1], Paduraru et al. [2] and Rashid et al. [3] address these issues by offering a decentralised, unchangeable ledger that improves data security and integrity. AIVs can safely communicate through vehicular ad hoc networks (VANETs) by combining blockchain technology with AI-driven prediction algorithms [4–6]. VANETs are essential for enhancing decision-making, lowering traffic, and boosting general road safety. The blockchain's cryptographic features guarantee the security of private user data and the immutability of data sent between cars.

According to recent research, AIV systems that integrate blockchain technology and artificial intelligence (AI) can improve security, decentralise data access, and guard against typical threats like Sybil attacks and data manipulation [1, 7, 8]. The potential contribution of blockchain technology to the security of autonomous intelligent vehicle networks is examined in this book chapter. It demonstrated how much it may increase the reliability of AI-based decision-making and secure communication networks. Blockchain has the ability to greatly aid in the creation of a secure and effective future for autonomous vehicles by tackling these crucial issues.

In this book chapter, we propose an architecture that will increase the security of autonomous intelligent vehicle communication systems by combining blockchain technology with AI-based prediction models, particularly reinforcement learning. By guarding against data manipulation and safeguarding the information transmitted between cars, this system will enhance decision-making in the AIV industry.

This book chapter aims to determine the significance of blockchain technology for autonomous intelligent vehicles, as well as the difficulties and potential integrations with various AIVs. The existing designs that have been offered will be analysed to ascertain the possible disadvantages and opportunities associated with integrating blockchain technology into an autonomous intelligent vehicle system.

The aims led to the following research questions:

i. Which fundamental ideas and characteristics of blockchain technology apply to intelligent, self-governing cars?
ii. What aspects of autonomous intelligent vehicles (AIVs) may blockchain technology help with?
iii. How may blockchain technology improve AIVs' efficacy, transparency, and security?
iv. How might the combination of AIVs and blockchain technology promote innovation and adoption?
v. What are the potential directions for future study?

The book chapter's precise objectives are as follows:

i. Give an overview of blockchain technology and its core concepts.
ii. Examine the key features of autonomous intelligent vehicles (AIVs).
iii. Analyse the benefits and challenges of using blockchain technology with AIVs.
iv. Discuss upcoming research and developments at the intersection of AIVs and blockchain.

To answer the aforementioned research objectives, the researchers suggested that blockchain technology can generate consistent, real-time, and up-to-date transactions without the need for a central authority. It will guarantee the creation of unquestionably secure vehicle and driver records for the course of the vehicle's usage lifecycle. Resolving issues with data integrity and reconciliation will result in a safe and secure vehicle information system. Simultaneously, the Blockchain-based vehicle information management system will acquire accurate and up-to-date data through integration with smart toolkits. It will give the stakeholders in the vast and interconnected car ecosystem—who do not trust one another—an effective, safe, and non-human intervention-free platform for data management and communication.

1.1 Overview of Blockchain Technology

All participating users maintain a Blockchain, which is a collection of interconnected blocks used to store enormous amounts of data. In a peer-to-peer network, BC (blockchain) allows all users to perform and validate transactions directly without the need for a central authority [9, 10]. Blocks are the simplest building block of a blockchain; they record every public transaction that is executed, validated, and subsequently saved.

The concept of blockchain was first proposed in 2008 [11] and introduced in early 2009. All committed transactions are kept in an expanding chain of blocks that function as a public ledger. Blockchain's decentralisation, auditability, permanence, and anonymity are among its key characteristics. Cryptographic hashes, asymmetric cryptography for digital signatures, and distributed consensus methods are some of the cutting-edge technologies used by blockchain. By facilitating the development of decentralised transactions, the application of these technologies lowers costs and increases efficiency [12].

The data on blockchains is kept in a connected series of blocks, each of which is secured by cryptographic techniques and connected to the one before it. Since the network's peers are connected via a P2P topology, there is not a single point of failure. Additionally, a consensus process is used by the network to guarantee that block sequences and transactions are widely accepted. This keeps data reliable and consistent across all nodes [13]. In the blocks that comprise this public ledger, each transaction on the blockchain is represented as a distinct record [14]. The second milestone in the blockchain's history started when smart contracts first appeared as a crucial component of the technology.

Blockchain 2.0 is the term used to describe this. Computer programs known as "smart contracts" operate within the blockchain network. Complex programmable transactions are made possible by these programs' ability to specify triggers, conditions, and business logic [15]. Long before the blockchain concept was created, computer scientist and cryptographer Nick Szabo was the first to refer to a smart contract by this moniker [16].

Bitcoin, the most well-known user of blockchain technology, is sometimes mistakenly thought of as being synonymous with blockchain. Nevertheless, blockchain technology is useful for more than just cryptocurrencies. Blockchain technology's primary feature is its capacity to enable transactions without the need for middlemen. This implies that it applies to a wide range of financial services, including remittances, digital assets, and online payments [17]. The variety of sectors that can profit from this cutting-edge technology has been greatly increased by smart contracts. Beyond digital banking, some emerging applications of blockchain include supply chain management, logistics, IoT, healthcare, personal identity systems, processing music royalties, anti-money laundering programs, voting, agriculture, and many more [18–20].

Permissioned and permissionless blockchains are the two primary varieties. Participation on a permissioned blockchain is limited to verified users whose identities and rights are controlled by a consortium or authority. More privacy and performance are provided by this method, but it necessitates confidence in the organisation that oversees membership. Permissioned blockchains include Hyperledger Fabric and R3 Corda. A permissionless blockchain, on the other hand, is open to the public, and anyone can take part in transaction validation. Although this open model—best represented by Bitcoin and Ethereum—promotes widespread decentralisation, it may have poorer throughput and less control over data visibility, which makes it appropriate for situations that are more open or untrustworthy.

Blockchain has the potential to transform a number of additional uses, including supply chains, the delivery of distant services like travel and tourism, the sale of art through digital media transmission, and cutting-edge platforms that move computation towards data sources and distributed credentialing [21]. Numerous applications have been envisioned for blockchain data systems. Scientists and financial experts expect major developments centred around this innovative concept–especially in the logistics sector [22]. This new technology has even been referred to as disruptive

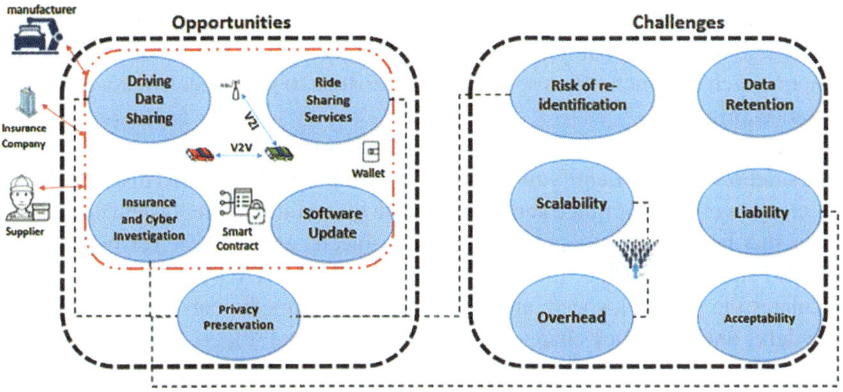

Fig. 1 Opportunities and challenges of integrating blockchain with CAV [24]

by certain authors. Numerous applications for the automotive sector and intelligent transport systems (ITS) have also been proposed, according to [23].

Blockchain technology has a lot of potential, but problems could still arise. Much like any new technology, it must overcome several challenges before it can be successfully adopted and applied in the automotive industry. These difficulties include those related to acceptance, scalability, laws, and technology. Understanding these difficulties is essential to maximizing blockchain technology's promise and minimizing its drawbacks.

Figure 1 shows the possibilities and difficulties of combining AIV and blockchain technology.

1.2 Evolution of AIV

Due to notable developments in perceptual technologies and processing capacities, autonomous driving technology is becoming more and more common in our daily lives [25]. The current skepticism over AIV safety, dependability, and societal impact is similar to the initial hesitancy and doubt surrounding the agricultural-to-industrial adaptation in the late eighteenth century. In light of the Industrial Revolution, AIV technology will eventually change our transport systems by integrating into our infrastructure, possibly more slowly than supporters expect.

Using a variety of sensors, including Lidar, radar, and cameras, autonomous vehicles (AVs) are at the forefront of technological advancement, precisely detecting and interpreting their surroundings. The journey towards completely autonomous transportation has advanced significantly with the integration of state-of-the-art technology, which enables AIVs to traverse complicated settings safely and effectively. There are six levels of driving automation: from Level 0 (completely manual driving) to Level 5 (full automation in all scenarios).

AIVs sense and navigate their surroundings using a variety of sensor technologies. These include global navigation satellite systems (GNSS), radar, LiDAR, and cameras, each of which offers special characteristics to guarantee effective and safe operation (Fig. 2).

In AIVs, RADAR (Radio Detection and Ranging) technology is essential because it uses radio waves to identify moving objects within predetermined distances of the vehicle, as well as their range and velocity. The transmitter in this system emits radio waves that bounce off moving or stationary objects, and the receiver picks up the reflected signals (see Fig. 3). In relation to the AIV, the RADAR system accurately calculates the position, speed, and direction of any impediments by examining the time delay and frequency change in the returned waves [27].

LiDAR is a remote sensing technique that measures distance, much like RADAR. LiDAR uses laser beams to measure distance by shining an invisible laser light on things that are visible to the human eye, as opposed to RADAR, which uses radio waves. These laser beams are subsequently reflected by the surroundings and sent to the device, where a photo detector records the signals that are returned. LiDAR calculates the accurate distance by measuring the time it takes for the laser to travel to and fro the target using the Time of Flight (ToF) method [29] (Li et al., 2020).

This capability enables LiDAR to collect millions of data points and produce detailed maps. Additionally, LiDAR continually scans 360 degrees to produce extensive 3D point clouds of its surroundings. Each pixel contains motion speed and depth data, giving users a thorough perspective of the surroundings. Figure 3 illustrates this.

In order to observe and comprehend the environment around the vehicle, cameras are essential in AIVs. They are essential for identifying traffic signs, pavement markings, and other dangers like obstructions. Generally speaking, cameras are reasonably priced, and with the right software, they can create high-resolution photographs of their surroundings and differentiate between fixed and moving objects in their range of vision [30]. AIVs can reliably and precisely navigate complicated situations, thanks to the integration of cameras and sophisticated processing algorithms.

Since real-time vehicle position, navigation, and timing are the fundamental features of any navigation system when paired with a digital map, GNSS is essential to the operation of AIVs [31]. GNSS satellites use radio transmissions to send their location and current clock information. A carrier wave plus a message modulated onto it via a particular modulation technique make up each GNSS signal. These days, GNSS primarily consists of four standard satellite navigation systems: the US-owned Global Positioning System (GPS); the EU-owned Galileo; the Russia-owned GLONASS; and the China-owned BeiDou. The most well-known of these systems is the GPS, which offers positioning, navigation, and timing (PNT) services to users.

To increase the precision of corrections in positioning data, AIVs now employ sophisticated methods such as Real-Time Kinematic (RTK) positioning and differential global positioning systems (DGPS) [31]. With the advent of GNSS-RTN systems, accurate positioning has been possible across distances ranging from 50 to 70 kms from a base station. It is advised that reference stations be spaced no more than 70 kms apart. In contrast to the 1 to 10 m precision usually observed without these sophisticated sensors, systems can attain positioning accuracy in the 1 to 5 cm range

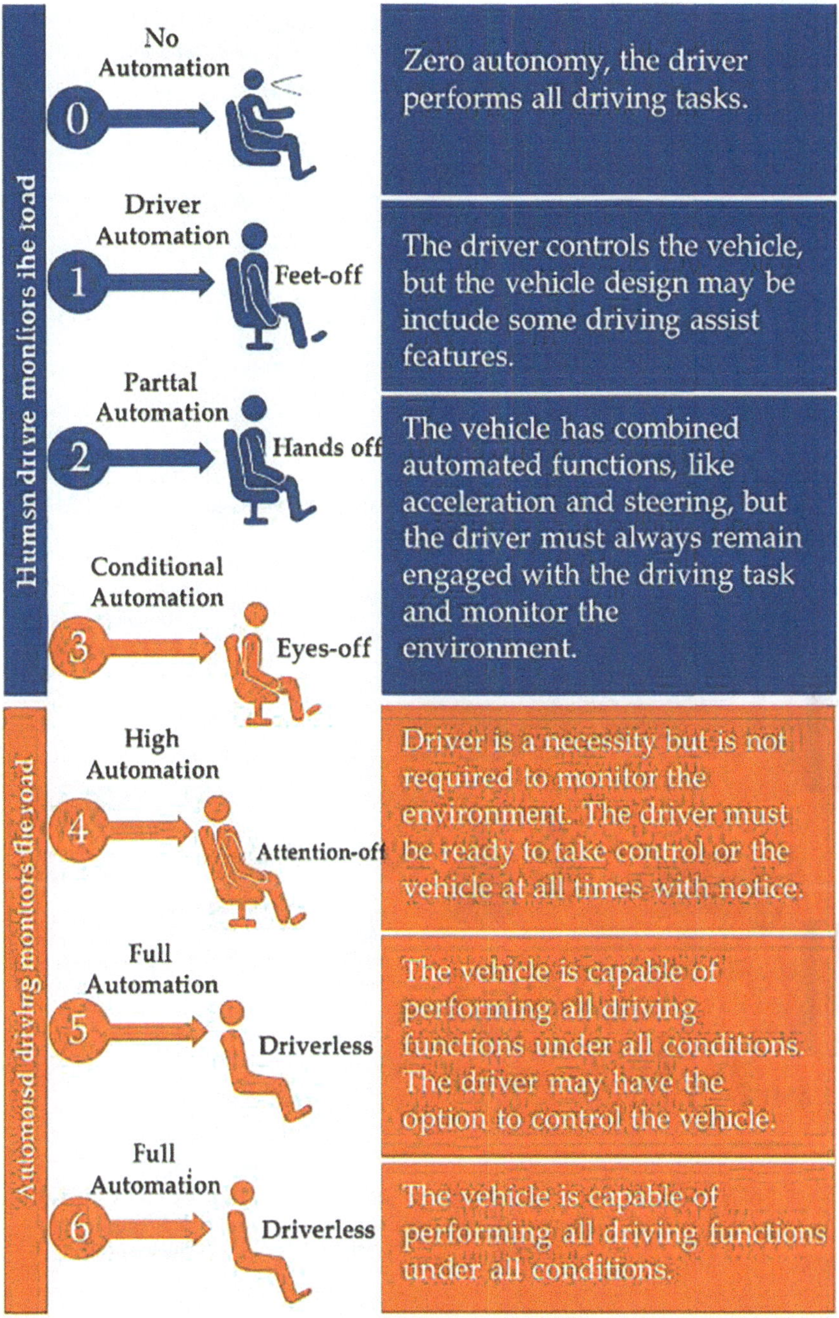

Fig. 2 Five levels' automation level description and driver support [26]

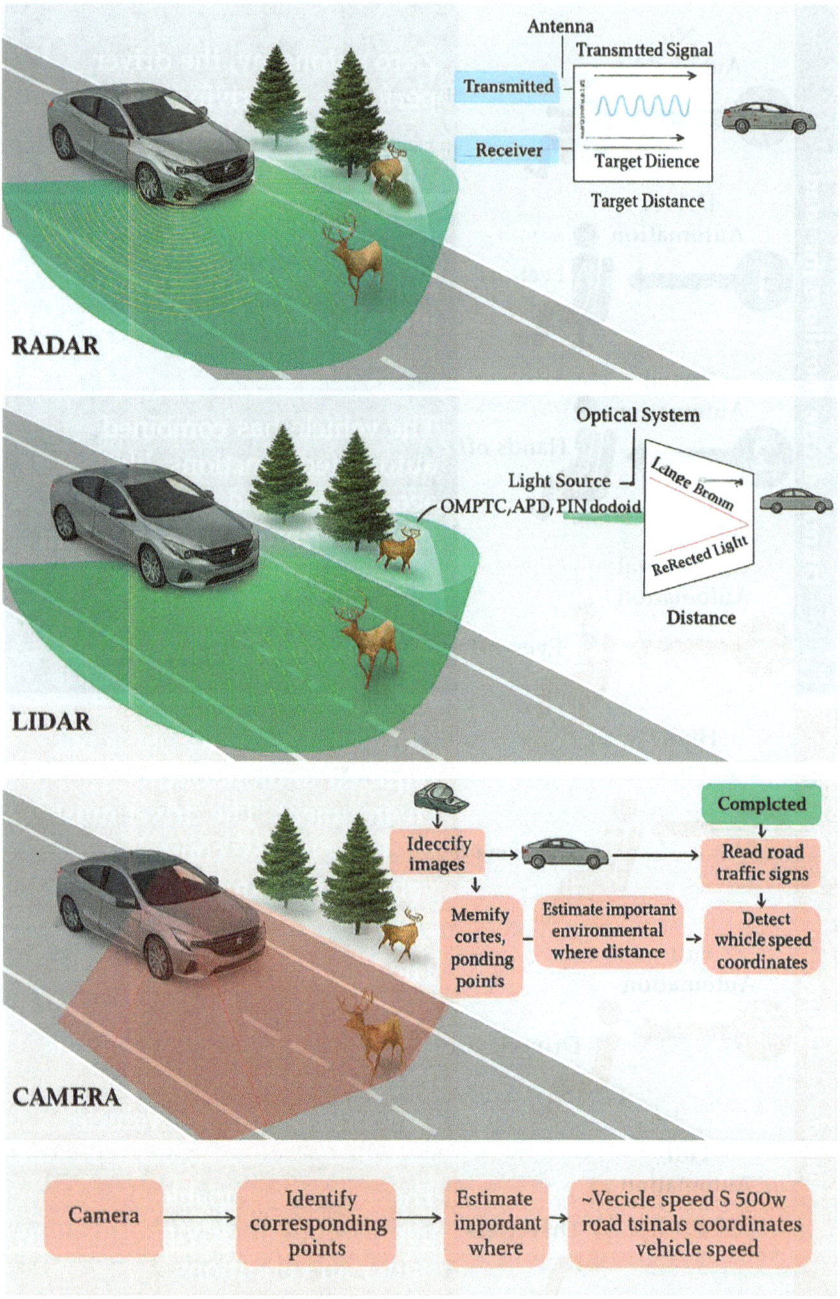

Fig. 3 The basic principle of sensors [28]

by employing GNSS-RTN [32]. The position error in DGPS might be as small as 1 cm. Positioning errors can be minimised to as little as 1 cm with DGPS [31].

Although all sensor types are appropriate, the perceptual accuracy and reliability of an AIV can be significantly increased by combining LiDAR, RADAR, and camera data. By combining the usage of several sensors, these systems are able to take into consideration the shortcomings of each sensor and provide a reliable framework for environmental variable detection and reaction. For instance, cameras might not work well in low visibility, but LiDAR and RADAR can still detect barriers and estimate distances. When combined, these technologies give autonomous cars the ability to constantly and precisely perceive their surroundings, which increases their safety and operational efficiency. These initiatives show enhanced capabilities that can greatly increase AIV accuracy and dependability through sensor fusion and integrated machine learning models.

1.3 Security Challenges in Autonomous Intelligent Vehicle Systems

There are several barriers and difficulties in the deployment of AIVs that keep these systems from being widely and successfully used. These issues stem from a variety of operational, legal, and technical concerns. AIV solutions are required to solve the current challenges to speed up technological advancement in the transportation sector, including:

i. Technical Interoperability Issues

   One of the primary problems with ITS (Information and Technology Systems) deployment is the lack of technical interoperability between various Cellular Vehicle-to-Everything (C-V2X) solutions and current technologies like LTE-V2X (Long Term Evolution-based Vehicle-to-Everything) and 5G-NR VANET (Fifth-Generation New Radio Vehicular Ad Hoc Networks). Large-scale implementation is made more challenging by the absence of a uniform standard that would enable the smooth integration of various systems and technologies [33].

   The incompatibility of different vehicle-to-everything (V2X) protocols, such as DSRC (IEEE 802.11p), LTE-V2X, and the more modern C-V2X standards, is a significant technical problem with AIV deployment. Despite sharing a 5.9 GHz spectrum, these protocols are incompatible at the communication level, which leads to issues with integration and fragmentation. Furthermore, there are differences in the infrastructure requirements: C-V2X can use cellular networks in addition to roadside devices, but DSRC needs specialised hardware only for short-range communications [34].

   To solve these problems, certain towns and corridors have started implementing dual-mode roadside units that handle both C-V2X and DSRC. Backward compatibility throughout the changeover period is made possible by this. Additionally, to communicate with older vehicles, more recent 5G-based C-V2X equipment will revert to LTE-V2X. Long-term interoperability. Successful ITS deployment depends on ongoing standardisation (such as IEEE 802.11bd, 5G NR V2X) and regulatory harmonisation, even while hybrid techniques provide a short-term fix [35].

ii. Issues with ITS Deployment and Scalability

The widespread adoption of AIV technologies raises a variety of complex challenges due to the huge number of stakeholders. This frequently results in disjointed programs that are challenging to implement in other locations. Since each plan is tailored to meet particular local demands, it is challenging to achieve the broad adoption needed to significantly alter urban mobility and sustainability [36].

iii. Privacy and Information Security in AIV

Because depends so heavily on real-time data interchange and communication, protecting user information has become a top responsibility. Establishing guidelines for data protection in information and technology systems (ITSs) requires the General Data Protection Regulation (GDPR). The goal of initiatives like CyberSec4Europe and Data Vaults is to develop safe frameworks for handling mobility data. Furthermore, the European Commission backs cybersecurity-by-design initiatives, making sure that intrusion detection systems, authentication protocols, and encryption are included in ITS platforms to guard against malicious attacks and data breaches [37].

iv. Confidentiality

Insurance information is protected against unauthorised access by maintaining confidentiality. One of the most crucial issues with blockchain and cloud computing is data secrecy, which ensures that only authorised users on the network may access data. The service provider must thus make sure that client data is not accessible to unauthorised people. A data owner typically loses control over their data when they share it with others through the cloud or blockchain, since they no longer have physical control over it. Multiple tasks are executed concurrently, and customers utilise shared resources in cloud computing. As a result, issues with secrecy and potential data leaks are raised.

To breach cloud confidentiality, attackers may employ techniques like packet eavesdropping, password attacks, port scanning, wiretapping, keylogging, and social engineering. At a certain point, blockchain technology can guarantee confidentiality. Due to the decentralised structure of the blockchain, the data pieces are transparent to everyone who shares them with others on a single blockchain.

Because of this, employing blockchain to maintain confidentiality is difficult. Blockchain's decentralised nature has significant ramifications; for instance, it makes it possible for anybody to view the transactions. As a result, businesses use cryptography to preserve blockchain integrity and secrecy. Blockchain technology uses a private/public key method to encrypt data and sign communications. Blockchain information can be used to spot scams and illicit transactions. As a result, blockchain can offer some degree of anonymity and successfully safeguard pseudonymity.

v. Immutability and Integrity

Data integrity is the process of preserving data and guaranteeing consistency. The service provider must guarantee data integrity by updating clients on the status of their stored data. The integrity of the cloud may be threatened by a Sybil attack, in which a hacker fabricates enough false identities to compromise the real system and seize control of the network. This can be accomplished by undermining current devices or by introducing new ones. Cloud providers employ an integrity management solution to engage a third party to guarantee data integrity. The data owner's readiness to provide a third party with data integrity verification, however, is a basic flaw in this strategy.

In addition to offering the standard services, cloud computing must guarantee platform users' dependability and quick access. Applications will not operate, and data will not be accessible if there is a server issue. Concerns regarding the dependability of cloud computing have been raised in recent years due to the rising frequency of network failures [38].

A blockchain's well-formed transactions, authentication, and auditing capabilities, as well as the P2P network's reliability, allow for the preservation of the data's integrity. Blockchain also uses a data integrity verification mechanism based on Merkle trees. The metadata must be altered before an attacker can attempt to alter the data that is stored on the blockchain. Given that each peer keeps the same data based on the addition of new blocks to the blockchain, attempts at subversion are impossible.

Due to the P2P network's dependability, its well-formed transactions, and its authentication and auditing capabilities, a blockchain can maintain the integrity of the data that has been saved. Blockchain also makes use of a Merkel-tree-based data integrity verification system. An attacker must alter the metadata before they may alter the data that is stored on the blockchain. The network is resistant to efforts at subversion since all peers keep the same data based on the addition of new blocks to the blockchain. If an open-source blockchain program has errors or flaws, the reliability of the data could be in danger. Many blockchains have serious vulnerabilities in their source code that, if exploited, might result in an inadvertent hard fork [39].

However, a blockchain is safer than the cloud since it is decentralised and dispersed. Fraud is far more difficult in blockchains than in cloud computing since there is not a single point of failure. However, when employing blockchain, performance and stability are traded off. In conclusion, a variety of reasons can lead to the compromise of the integrity of insurance data, whether it is stored in the cloud or on a blockchain: (1) the quantity of gadgets (2) the quantity of players (cloud service providers, insurance companies, security agencies, and data owners); and (3) the dynamic data sources.

vi. Accessibility

The word "availability" refers to the fact that information is accessible when required. For a variety of stakeholders, such as law enforcement agencies, insurance companies, and data owners, data availability is essential. However, dishonest actors might abuse the same data to create serious ethical and social issues. For insurers, DoS attacks, data exfiltration, malware infections, financial transaction theft, zero-day exploits, and phishing emails are the most common types of cyber-attacks [40]. These hacks primarily affect insurers by causing business interruptions and substantial costs for both the company and its clients. One of the most widely utilised techniques by attackers to lower system availability is distributed denial of service. This attack's primary goal is to bring a system down.

vii. Privacy

Data privacy, as defined by the General Data Protection Regulation (GDPR), is the extent to which people have control over who can access their personal information. For many apps and organisations, privacy is a major concern [41]. Data privacy is a major concern for both clients and insurance firms. There are no explicit guidelines regarding data ownership in the insurance sector. Even though a driver's record may belong to them alone, a police officer or insurance inspector may also want to see it.

Insurance firms are motivated by the privacy challenge to develop policies and guidelines that delineate ownership limits. It should be possible to detect and reduce adverse privacy consequences, and decision-makers should be aware of these so they can take the necessary action. For cloud computing, privacy is a crucial issue that needs to be given top priority in order to maintain regulatory compliance and user confidence. It can be difficult to protect auto records from cloud service providers, one another, and other insurance companies. All insurance companies need certain access to records in order to safeguard their own data.

The identity of a car is typically linked to its owner due to legal and insurance concerns. Consequently, tracking a car may regularly violate its owner's privacy. Given that the accident data that insurance firms use is based on moving automobiles, there are significant and particular privacy concerns. Investigators may encounter privacy issues when insurance companies use video analysis for forensic investigations. Analysing the video should be done without violating the organization's or the customer's privacy.

With or without authorisation, data can be accessed in a blockchain. The security, validity, and integrity of a private blockchain are typically guaranteed by a regulatory body. In a public blockchain, most independent peers protect the integrity and security of the network. Sensitive information is stored in the blockchain's chain ledger as transactions that authorised users can view and share. Traceable hints can reveal identities and other related information, even though blockchain transactions can be made anonymously [42–44]. Third-party programs that monitor users' many accounts can be compromised and subverted, and transactions can be connected to IP addresses to uncover even more information about a person.

For a blockchain network and its applications to function, massive volumes of data must be transferred in real time. Therefore, maintaining user anonymity if a malevolent actor has access to all transaction data within the blockchain is the next most crucial problem in blockchain technology [42]. Transaction leakage is a significant issue with blockchains since block data is visible to all network users. One of the most important concerns in the auto-insurance industry's use of vehicle and driving data is protecting the privacy of the driver.

viii. Access Control

One security measure that limits access to specific resources in a computer environment is access control. Following the issuance of a user's credentials and authorisation, this basic security rule controls access to resources and reduces organisational risk. It keeps unauthorised persons from accessing the system. Due to the wide variety and growing number of dispersed devices, there are several access control problems in AIV-related data.

Therefore, before using any system, service, or device, identification is always required. Since cloud-based auto insurance finds it difficult to provide secure network file storage and access management, data access is becoming more and more important as cloud computing grows [45]. Among the security models that are now available, there is no ideal one for access control [46].

The challenges with access control in cloud computing are related to privacy and availability. Unauthorised access leads to a loss of privacy, and access denied to a legitimate user result in a loss of availability. Moreover, an unauthorised person who gains access to the control network has the ability to alter data in ways that jeopardise data integrity, learn the communication protocol, and initiate various assaults [46]. Given that it directly affects all security principles, including availability, privacy, and integrity, access control is therefore the most important obstacle.

ix. Storage

Storage is the term used to describe digitally stored data, files, and documents for later use. A storage system can use optical, electromagnetics, or other media to safeguard and recover data when needed. Cloud storage provides dependable, easy-to-use, and reasonably priced storage. Insurance companies and insurers can swiftly access data due to cloud computing. They can utilise the data and trends to produce creative solutions that satisfy their customers' needs [47]. Despite the many benefits of cloud storage, insurance firms still have to deal with a number of issues including loss of control, reliance on the internet, and even migration.

Blockchain technology offers immutable storage, allowing transactions to be added but never changed or deleted. Large amounts of data cannot be efficiently stored on a blockchain due to replication across all nodes. Additionally, the computational difficulty of running blockchain nodes rises with the amount of data processed, transferred, and saved. In other words, the cost model for data storage in a blockchain differs from that of traditional data storage.

1.3.1 Other Difficulties

1. Making Use of Energy

   Enough energy is required to supply dependable electricity to the hundreds of computers in data centres. Moreover, long-distance data transmission uses more energy than local storage. Cloud computing consumed 30 billion watts of electricity per year, according to a 2012 study [48].

   Since complicated calculations need a lot of processing power from the participating computers, many blockchain solutions validate transactions using a proof-of-work consensus mechanism. Blockchain relies on hundreds of miners adding and verifying transactions with powerful electric equipment. The number of miners operating determines how much energy is used. Consequently, blockchain alone uses 204.5 terawatt-hours of electricity every year. Bitcoin currently uses more electricity than Argentina, a country of 45 million people, with an annual usage of about 150 terawatt-hours [49].

2. Dependency on outside sources

   Most blockchain-related systems have infrastructure problems because the technology is still in its infancy. Due to their reliance on outside tools and technology, the present approaches to development, testing, and performance evaluation raise concerns about their decentralized nature. Insurance businesses and data owners are wary of relying on blockchain and cloud computing for data security and privacy because of the use of third-party storage.

3. Scalability

   The growing number of nodes and transactions on the blockchain has led to scalability problems. This problem is prevalent in public blockchain systems since every node needs to retain information and carry out calculations in order to validate every transaction. As a result, public blockchains constantly need a lot of storage space, minimal latency, and processing power.

1.4 How Blockchain Can Help Improve Security

The safety and dependability of vehicle communication systems are crucial as Autonomous Intelligent Vehicles (AIVs) become a more significant component of contemporary transportation. Data flow between autonomous cars and urban infrastructure depends on interconnected networks, which present serious dangers like impersonation, data tampering, and cyber-attacks. These weaknesses have the potential to compromise AIV security and decision-making accuracy in real-time traffic situations.

A decentralised, unchangeable ledger that improves data security and integrity is provided by blockchain technology, which presents a viable answer to these problems [1–3]. AIVs are essential for lowering traffic congestion, enhancing decision-making, and raising overall road safety since they can securely communicate by combining blockchain technology with AI-driven prediction algorithms. The blockchain's cryptographic features guarantee that private user data is kept safe and that data transferred between cars cannot be altered.

1.5 AIV Security Challenges Versus Blockchain Solutions

The introduction of Autonomous Intelligent Vehicles (AIVs) holds up the prospect of numerous advantages, such as improved safety and decreased energy use, pollution, and traffic. However, if left unchecked, these engines' numerous security and privacy flaws could undermine the anticipated advantages. AIVs will give hackers new ways to launch destructive assaults, which is a danger to data security and mobility in the future.

Blockchain technology could provide AIVs with robust defence against malicious attacks. Although blockchain and AIVs operate differently, they can complement one another and resolve many of the AIVs' security and privacy concerns when combined.

Table 1 presents various AIV security challenges versus blockchain Solutions.

1.6 Sensor Vulnerabilities Versus Blockchain Support

This section demonstrates the risks and weaknesses to which sensors may be subject as well as the ways in which blockchain impacts them. This is displayed in Table 2.

1.7 Blockchain Trials in Transportation

The implementation of blockchain technology can have a number of beneficial effects on transportation management. Blockchain technology is seen as potentially disruptive for transport systems in the very well-written report of Catapult Transport Systems [50]. However, this report also highlights the fact that many of the experts who participated in a survey on BT (Blockchain Technology) trends appear to be poorly informed about this new system.

Table 1 AIV security challenges versus blockchain solutions

S/N	Challenge	Explanation	Solution
1	Data privacy concerns	Due to its requirement for massive datasets, AI raises questions regarding data security and possible breaches	The decentralized design and cryptography capabilities of blockchain
2	Malicious assaults	It is possible to manipulate AI models by providing them with false information	Put strong security measures in place and train AI models securely
3	Problems with scalability	Both blockchain and artificial intelligence have scaling issues	Use sharing, rollups, or layer 2 scaling options
4	Challenges of interoperability	Blockchain and AI integration calls for rigorous preparation	Create standardized protocols and APIs
5	AI model tampering	Modification of AI models without authorization	The decentralized design and cryptographic features of Blockchain
6	Cybersecurity risks	Numerous cyberthreats to blockchain and AI systems	Bringing together blockchain-secured data storage and AI-driven danger identification

Table 2 Sensor vulnerabilities versus blockchain support

S/N	Sensor vulnerability	Description	Blockchain support
1	Data manipulation	Unauthorized modification of sensor information	Unchangeable ledger and cryptographic features
2	Attacks that cause denial of service	Sensors overflowing with requests	Distributed networks and decentralized architecture
3	Loss of data	Data loss from sensors as a result of a hardware malfunction	Unchangeable ledger and decentralized data storage
4	Unauthorized entry	Unauthorized entry because of sensor information	capabilities of cryptography and an access control system
5	Interception of data	Intercepting sensor data while it's being transmitted	Secure protocols for communication and end-to-end encryption
6	The spooling of sensors	Injecting false sensor data into the system	Immutable ledger and methods for validating data

The final section of the work discusses how logistics and transport may change in the future due to the convergence of cutting-edge services and technologies like 5G (the next wireless mobile data standard), Mobility as a Service (MaaS), the Internet of Things (IoT), deep learning in Artificial Intelligence (AI), and scattered smart devices.

When we discuss transportation, we mean the movement of people or objects from one place to another. Everything related to moving goods will be separated out in the following and considered a part of a study cluster on supply chains and logistics. We will illustrate that blockchain-based systems are being researched in the transportation industry, specifically with regard to the concept of new autonomous and connected vehicles.

Numerous positive effects on transportation management can result from the application of blockchain technology. According to Catapult Transport infrastructure's extremely well-written report [21], blockchain technology has the potential to disrupt transport infrastructure. This report also draws attention to the fact that a large number of the experts who took part in a survey on BT trends seem to lack sufficient knowledge about this new system. In closing, this study demonstrates how the convergence of advanced technologies and services such as Mobility as a Service (MaaS), the Internet of Things (IoT), deep learning-based artificial intelligence (AI), 5G (the next generation of wireless mobile data protocol), and distributed smart objects may change the logistics and transportation landscape of the future.

Blockchain technology can be applied to many AIV applications, such as decentralised marketplaces for transportation services, smart contracts for driverless cars, and safe data transmission between automobiles, infrastructure, and service providers. AIV uses blockchain technology for platooning, which is the process of arranging several vehicles in a close-knit, coordinated configuration. A method of effective vehicle convoy control that has emerged as a solution to serious traffic problems like air pollution, traffic safety hazards, and congestion is the platooning formation [51].

Due to the reduced but controlled spacing between vehicles and guaranteed constant speeds, the practice of arranging groups of cars to go in tight formation improves traffic flow by allowing more cars on the road at once without creating additional congestion.

Secure vehicle identification systems are another example of how blockchain technology is being used in AIV. In order to improve autonomous driving skills and enable new features like smart traffic management or technologies designed to prevent or lessen traffic accidents, it is necessary to create secure vehicle communication. By facilitating secure communication and dynamic key generation without prior key exchanges, blockchain technology and identity-based cryptography simplify encryption and make it incredibly scalable and effective.

Lastly, the process by which an insurance company or other accountable entity responds to and manages claims for compensation from policyholders who have experienced losses or damage is known as insurance claim management. This procedure comprises several steps, such as claim notification, damage assessment, compensation settlement, and, if required, dispute resolution. In addition to minimising the financial impact on the insurer, preserving policyholders' faith requires fair, open, and prompt claims handling. Technological advancements like automation, artificial intelligence, and blockchain technology are having a beneficial impact on this industry by facilitating quicker and more efficient claims processing, which helps to reduce fraud.

2 Review of Related Works

The present study was carried out using Kitchenham's technique, one of the most widely used and effective Systematic Literature Review (SLR) approaches in software engineering [52]. This technique offers rigorous processes for analysing research knowledge using a methodology that is reliable and auditable. However, other authors have questioned and/or proposed modifications to Kitchenham's approach (de Silva et al., 2011). Kitchenham published an updated version of her technique in Kitchenham et al. (2010) in response to these critiques and recommendations for improvements. This SLR adheres to the most recent version of Kitchenham's model, as was previously mentioned. There are three steps to doing a systematic review.

i. Planning describes components such as the research topics, the evaluation process, and the necessity of the study.
ii. Conducting: this entails adhering to the established protocol.
iii. Reporting, which offers the final interpretation of every research question.

3 Method

This review study employed the content analysis method. From the various scientific databases, pertinent studies were retrieved. Web of Science, IEEE Explore, SpringerLink, Google Scholar, and Science Direct are a few of the databases that were searched. The following terms or phrases were used to find publications for this study: "Autonomous Intelligent Vehicles and Blockchain," "Overview of Blockchain Technology," "Autonomous Systems Security Challenges," and "The Role of Blockchain in Enhancing Security." This helped us focus our search on studies that solely relate to blockchain applications in AIV systems. Once the publications were sorted, they were examined and used in this review.

4 Result

In order to satisfy the study objectives and provide readers with accurate and sufficient information, the book chapter's results will be presented in a number of sections using tables and figures.

5 Discussion

Blockchain technology and its potential impact on driverless intelligent cars are covered by Alanazi [53], who focused on Intelligent Transportation Systems (ITS). Despite some challenges, blockchain technology is an amazing tool for improving the efficacy and security of Intelligent Autonomous Vehicles while simultaneously addressing concerns like security, privacy, and interoperability. In order to explore the acknowledged benefits and potential applications of integrating Blockchain technology into AIV, this systematic review (SLR) literature study looks at over 30 publications.

This study shows how smart city transit systems could undergo a radical change due to blockchain technology. To ensure maximum accuracy and efficacy, Rakhmanov and Wiseman [54] also covered the usage of data compression techniques to supply GNSS data to satellites and autonomous vehicles. They came up with a way to make the data smaller.

Additionally important and not to be overlooked are data security and storage. The multiplicity of data sources, their inconsistent behaviour, and the lack of confidence among various stakeholders are the main issues facing the car industry. To identify security flaws and improve security, it is critical to fully understand blockchain technologies. Although the auto insurance industry is still in the early stages of implementing these strategies, there is a chance to increase the efficacy and efficiency of Autonomous Intelligent Vehicles.

Integrating AI, big data, and IoT technologies with cloud and blockchain-based auto systems may enhance insurance administration and successfully detect and prevent insurance fraud. By accurately recording the timing, location, and other specifics of an accident, blockchain technology can protect insurance firms against fraud. AI can speed up the claims procedure and enhance customer service. Data monitoring and analysis are made easier by big data technologies. Blockchain technology can be used to track driving logs, operating data, and other relevant information about a vehicle.

Acknowledgements The authors genuinely appreciate the anonymous reviewers' insightful criticism and recommendations for enhancing the manuscript's quality.

References

1. Chen, L., Zhao, Y., Wang, M.: A secure and efficient blockchain-based system for vehicle-to-everything (V2X) communication. IEEE Trans. Intell. Transp. Syst. **23**(1), 1–14 (2022)
2. Paduraru, C., Jianu, A., Stefanescu, A.: Blockchain for artificial intelligence: an industry and literature survey. In: Proceedings of the 18th International Conference on Software Technologies (ICSOFT), pp. 135–150. SciTePress (2023)
3. Rashid, M., Alam, S., Khan, M.: Enhancing security in VANETs with blockchain and PBFT consensus mechanism. Comput. Netw. **184**, 107644 (2021)

4. Ramasamy, K., Subramanian, D.: Blockchain-based privacy solutions in vehicular ad-hoc networks (VANETs). Comput. Secur. J. **108**, 102456 (2022)
5. Smith, A., Jones, K., Lee, D.: Blockchain and smart contract applications in transportation systems. Transp. Sci. **54**, 130–144 (2023)
6. Zhu, H., Huang, X., Zhang, W.: Integration of blockchain and AI in autonomous vehicle safety. Int. J. Distrib. Sens. Netw. **19**, 1–12 (2023)
7. Park, D., Shin, J., Lee, Y.: Mitigating Sybil attacks in AV communication networks via blockchain. IEEE Veh. Technol. J. **13**, 51–63 (2023)
8. Patel, S., Kumar, A., Verma, P.: Implementation of smart contracts in AV communication for authentication and privacy. J. Automot. Secur. **48**, 102–118 (2023)
9. Singh, P., Elmi, Z., Lau, Y.Y., Borowska-Stefanska, M., Wisniewski, S., Dulebenets, M.A.: Blockchain and AI technology convergence: applications in transportation systems. Veh. Commun. **38**, 100521 (2022)
10. Abbas, S., Talib, M.A., Ahmed, A., Khan, F., Ahmad, S., Kim, D.H.: Blockchain-based authentication in internet of vehicles: a survey. Sensors **21**, 7927 (2021)
11. Nakamoto, S.: Bitcoin: A peer-to-peer electronic cash system (2008). Retrieved from https://bitcoin.org/bitcoin.pdf
12. Zheng, Z., Xie, S., Dai, H.N., Chen, X., Wang, H.: Blockchain challenges and opportunities: a survey. Int. J. Web Grid Serv. **14**(4), 352–375 (2018)
13. Yuan, Y., Wang, F.-Y.: Blockchain and cryptocurrencies: model, techniques, and applications. IEEE Trans. Syst., Man, Cybern. Syst. **48**(9), 1421–1428 (2018)
14. Tschorsch, F., Scheuermann, B.: Bitcoin and beyond: a technical survey on decentralized digital currencies. IEEE Commun. Surv. Tutor. **18**(3), 2084–2123 (2016)
15. Xu, X., Pautasso, C., Zhu, L., Gramoli, V., Ponomarev, A., Tran, A.B., Chen, S.: The blockchain as a software connector. In: 2016 13th Working IEEE/IFIP Conference on Software Architecture (WICSA), pp. 182–191. IEEE (2016)
16. Saputri, F.A.: Regulating the use of smart contract in Indonesia. JHK J. Huk. Dan Keadilan **1**, 42–50 (2024)
17. Peters, G., Panayi, E., Chapelle, A.: Trends in cryptocurrencies and blockchain technologies: a monetary theory and regulation perspective. J. Financ. Perspect. **3**, 1–25 (2015)
18. Christidis, K., Devetsikiotis, M.: Blockchains and smart contracts for the internet of things. IEEE Access **4**, 2292–2303 (2016)
19. Manski, S.: Building the blockchain world: technological commonwealth or just more of the same? Strateg. Chang. **26**(5), 511–522 (2017)
20. Monrat, A.A., Schelén, O., Andersson, K.: A survey of blockchain from the perspectives of applications, challenges, and opportunities. IEEE Access **7**, 117134–117151 (2019)
21. Casino, F., Dasaklis, T.K., Patsakis, C.: A systematic literature review of blockchain-based applications: current status, classification, and open issues. Telematics Inform. **36**, 55–81 (2019)
22. Friedlmaier, M., Tumasjan, A., Welpe, I.M.: Disrupting industries with blockchain: the industry, venture capital funding, and regional distribution of blockchain ventures. In: Proceedings of the 51st Hawaii International Conference on System Sciences, pp. 1–10 (2018)
23. Yuan, Y., Wang, F.-Y.: Towards blockchain-based intelligent transportation systems. In: 2016 IEEE 19th International Conference on Intelligent Transportation Systems (ITSC), pp. 2663–2668. IEEE (2016)
24. Dargahi, T., Ahmadvand, H., Alraja, M.N., Yu, C.-M.: Integration of blockchain with connected and autonomous vehicles: vision and challenge. **1**(April 2021), 9 (2021). https://doi.org/10.1145/3460003
25. Guanetti, J., Kim, Y., Borrelli, F.: Control of connected and automated vehicles: state of the art and future challenges. Annu. Rev. Control. **45**, 18–40 (2018)
26. Vargas, J., Alsweiss, S., Toker, O., Razdan, R., Santos, J.: An overview of autonomous vehicles sensors and their vulnerability to weather conditions. Sensors **21**, 5397 (2021)
27. Zhang, Y., Carballo, A., Yang, H., Takeda, K.: Perception and sensing for autonomous vehicles under adverse weather conditions: a survey. ISPRS J. Photogramm. Remote. Sens. **196**, 146–177 (2023)

28. Beigi, S.A., Park, B.B.: Impact of critical situations on autonomous vehicles and strategies for improvement. Futur. Transp. **5**, 39 (2025). https://doi.org/10.3390/futuretransp5020039
29. Royo, S., Ballesta-Garcia, M.: An overview of lidar imaging systems for autonomous vehicles. Appl. Sci. **9**, 4093 (2019)
30. Ignatious, H.A., El Sayed, H., Khan, M.: An overview of sensors in autonomous vehicles. Procedia Comput. Sci. **198**, 736–741 (2021)
31. Dasgupta, S., Rahman, M., Islam, M., Chowdhury, M.: A sensor fusion-based GNSS spoofing attack detection framework for autonomous vehicles. IEEE Trans. Intell. Transp. Syst. **23**, 23559–23570 (2022)
32. Raza, S., Al-Kaisy, A., Teixeira, R., Meyer, B.: The role of GNSS-RTN in transportation applications. Encyclopedia **2**, 1237–1245 (2022)
33. Essawy, S.K.E.G.: Achieving Interoperability in Mobility as a Service: A Data Ecosystem Leveraging Semantic Web Technologies. Doctoral dissertation, University of Nottingham Malaysia (2024). Retrieved from https://core.ac.uk/download/613841379.pdf
34. Choi, J., Marojevic, V., Labib, M., Kabra, S., Rao, J., Das, S., Reed, J.H., Dietrich, C.B.: Regulatory options and technical challenges for the 5.9 GHz spectrum: survey and analysis. arXiv:1810.04846 (2018)
35. Khattak, A.J., Harris, A., Sartipi, M., Mahdinia, I., Moradloo, N., SafariTaherkhani, M.: Connected and automated vehicles investment and smart infrastructure in Tennessee Part 3: Infrastructure and vehicular communications: From dedicated short-range communications to cellular vehicle-to-everything (2023). arXiv:2304.02885
36. Psara, K., Papadimitriou, C., Efstratiadi, M., Tsakanikas, S., Papadopoulos, P., Tobin, P.: European energy regulatory, socioeconomic, and organizational aspects: an analysis of barriers related to data-driven services across electricity sectors. Energies **15**, 2197 (2022)
37. Cyber Security Network of Competence Centres for Europe (2019). Retrieved from https://cordis.europa.eu/project/id/830929
38. Godfrey, M., Zulkernine, M.: Preventing cache-based side-channel attacks in a cloud environment. IEEE Trans. Cloud Comput. **2**(4), 395–408 (2014)
39. Why the evolution of blockchain reliability is critical to protecting your digital assets: World Economic Forum (2020). Retrieved from https://www.weforum.org/agenda/2020/06/evolution-of-blockchain-reliability-and-digital-asset-protection
40. EU: Cyber risk for insurers: Challenges and opportunities. Publications Office of the European Union, Brussels, Belgium (2019)
41. Castagna, R., Lavery, T.: General Data Protection Regulation (GDPR). TechTarget (2022). Retrieved from https://www.techtarget.com/whatis/definition/General-Data-Protection-Regulation-GDPR
42. Blockchain security issues and how to prevent them. Fast Company (2022). Retrieved from https://www.fastcompany.com/90722111/5-blockchain-security-issues-and-how-to-prevent-them
43. Di Francesco Maesa, D., Mori, P., Ricci, L.: Blockchain-based access control. In: IFIP International Conference on Distributed Applications and Interoperable Systems, pp. 206–220. Springer (2017)
44. Steichen, M., Fiz, B., Norvill, R., Shbair, W., State, R.: Blockchain-based, decentralized access control for IPFS. In: 2018 IEEE International Conference on Internet of Things (iThings), GreenCom, CPSCom, and SmartData, pp. 1499–1506. IEEE (2018)
45. Alansari, S., Paci, F., Sassone, V.: A distributed access control system for cloud federations. In: 2017 IEEE 37th International Conference on Distributed Computing Systems (ICDCS), pp. 2131–2136. IEEE (2017)
46. Djenouri, Y., Yazidi, A., Srivastava, G., Lin, J.C.-W.: Blockchain: applications, challenges, and opportunities in consumer electronics. In: IEEE Consumer Electronics Magazine (2023)
47. Jabar, H.Y., Kumar Saini, D.: Cloud computing and accident handling systems. Int. J. Comput. Appl. **63**, 19–25 (2013)
48. Brdar, D.: Cloud computing energy's efficiency problems. NASDAQ (2022). Retrieved from https://www.nasdaq.com/articles/cloud-computing-energys-efficiency-problems

49. Feign, A.: How much energy does Bitcoin use? CoinDesk (2021). Retrieved from https://www.coindesk.com/business/2021/08/18/how-much-energy-does-bitcoin-use
50. Carter, C., Koh, L.: Blockchain Disruption in Transport: Are You Decentralized Yet? Catapult Transport Systems, Milton Keynes, UK (2018)
51. Fakhfakh, F., Tounsi, M., Mosbah, M.: Vehicle platooning systems: review, classification and validation strategies. Int. J. Netw. Distrib. Comput. **8**, 203–213 (2020)
52. Brereton, P., Kitchenham, B.A., Budgen, D., Turner, M., Khalil, M.: Lessons from applying the systematic literature review process within the software engineering domain. J. Syst. Softw. **80**, 571–583 (2007)
53. Alanazi, F.: The future of transportation: blockchain-powered solutions. Transp. J. **64**(1), 10 (2024)
54. Rakhmanov, A., Wiseman, Y.: Compression of GNSS data with the aim of speeding up communication to autonomous vehicles. Remote. Sens. **15**, 2165 (2023)
55. Chandrasekaran, K., Thomas, V.M.: Distributed access control in cloud computing systems. In: Encyclopedia of Cloud Computing, pp. 417–423 (2016)

Blockchain Fundamentals in Autonomous Intelligent Vehicle

Elvin Eziama, Diovu Chidiebere, Chukwudi E. Agbaraji, Ugboaja Samuel Gregory, Christian Nduka Onyeukwu, David A. Oduro, Benedict O. Ibe, Victor Eghujovbo, Ifeanyi F. Nwosu, Ogechi O. Onyebuchibe, JohnPaul Adimonyemma, Harrison Osasogie Edokpolor, and Dan Ali

Abstract Blockchain can provide trust, transparency and resilience in AIV networks, and its performance relies on the selection of consensus mechanism. Conventional solutions, e.g., PoW, PoS, are not feasible in VANETs since they consume a high energy that resulted in a long wait and in the dependence on incentive in Cryptocurrency. Widely deployed canonical ones, such as BFT-based permissioned protocols, tend to achieve faster finality and achieve fault tolerance, but also suffer on scalability. Reputation-based solutions like Proof-of-Reputation (PoR) increase the efficiency of consensus based on the trust of nodes and minimize the risk posed by malicious nodes. New alternatives such as ledgers based on directed acyclic graphs

E. Eziama (✉) · V. Eghujovbo
Mechanical, Automotive and Material Science, University of Windsor, Windsor, ON, Canada
e-mail: eziama@uwindsor.ca

V. Eghujovbo
e-mail: eghujov@uwindsor.ca

D. Chidiebere
Biomedical Engineering Department, Federal University of Allied Health Sciences, Enugu, Enugu, Nigeria

C. E. Agbaraji
Department of Computer Science, Bishop's University, Sherbrooke, QC, Canada

U. S. Gregory
Department of Computer Science, Michael Okpara University of Agriculture, Umuahia, Abia, Nigeria
e-mail: ugboaja.samuel@mouau.edu.ng

C. N. Onyeukwu
Department of Computer Science, Bishop's University, Sherbrooke, QC, Canada
e-mail: onyeukwu@cs.ubishops.ca

D. A. Oduro
Senior FP&A Analyst, Gulf of Maine Research Institute, Portland, ME, USA

Elvin Eziama, Diovu Chidiebere, Chukwudi E. Agbaraji, Ugboaja Samuel Gregory, Christian Nduka Onyeukwu, David A. Oduro, Benedict O. Ibe, Victor Eghujovbo, Ifeanyi F. Nwosu, Ogechi O. Onyebuchibe, JohnPaul Adimonyemma, Harrison Osasogie Edokpolor, and Dan Ali, These authors contributed equally to this work.

S. O. Ajakwe and D.-S. Kim (eds.), *Blockchain for Security of Autonomous Intelligent Vehicles*, Studies in Big Data 194,
https://doi.org/10.1007/978-3-032-20851-4_3

(DAG) have emerged that claim to be able to offer higher throughput; however, they also have unresolved security issues. In general, no single consensus solution describes all AIV scenarios, and hybrid and hierarchical solutions are probably necessary to achieve a trade-off between efficiency, resilience, and real-time safety.

Keywords Blockchain · Autonomous and intelligent vehicles (AIV) · Consensus mechanisms · Byzantine fault tolerance (BFT) · Proof-of-reputation (PoR) · Directed acyclic graph (DAG) · Vehicular networks · Security · Scalability · Latency

1 Introduction

The rapid evolution of Autonomous Intelligent Vehicles (AIVs) offers the potential to revolutionize transportation with improvements in safety, efficiency, and mobility. Fueling these cyber-physical systems is a multifaceted mix of sensors, artificial intelligence (AI), and pervasive connectivity—centering on the connectivity technology, Vehicle—to—Everything (V2X) which "sees" our world, and takes critical action with little human involvement. However, such autonomy and connectivity presents deep challenges for security, data integrity, and trust among a diverse group of participants (vehicles, infrastructure, service providers) that can even be unknown or potentially adversarial [1].

The security foundation of current Intelligent Transportation Systems (ITS) is typically based on a centralized or federation architecture including, for example, the classic Public Key Infrastructure (PKI) for credential management. These models show some key weaknesses, such as single point of failure, scalability bottleneck,

B. O. Ibe · I. F. Nwosu · O. O. Onyebuchibe · D. Ali
School of Science, Engineering and Environment, The University of Salford, Salford, Greater Manchester, UK
e-mail: B.o.ibe@edu.salford.ac.uk

I. F. Nwosu
e-mail: f.i.nwosu@edu.salford.ac.uk

O. O. Onyebuchibe
e-mail: f.i.nwosu@edu.salford.ac.uk

D. Ali
e-mail: a.i.dan@edu.salford.ac.uk

J. Adimonyemma
Department of Industrial and Manufacturing Engineering, FAMU-FSU College of Engineering, Florida Agricultural and Mechanical University, Tallahassee, FL, USA
e-mail: johnpaul1.adimonyemm@famu.edu

H. O. Edokpolor
University of Bath, Bath, UK
e-mail: hoe23@bath.ac.uk

long delay in certificate revocation, which make them suffer from advanced cyber-attacks, e.g., DDoS (distributed denial-of-service) attack, MitM (Man-in-the-Middle) attack (Wang et al. 2020; Rigazzi 2017). As a result, the initial trust that the base is scalable, tolerant, trustworthy trust models in the deployment of AIVs that is smooth continues to be the key challenge.

In this context, blockchain (or distributed ledger technology (DLT) in general) makes a natural choice as a disruptive answer to such trust issues. Originally developed for cryptocurrency applications [2], blockchain decentralization, cryptographic one-way hashing, and consensus-driven validation provide a new model for AIV network trust management. By providing an immutable, distributed, tamper-evident ledger of transactions and events, the blockchain can cut down on the risks of a centralized authority. Offers a reliable and transparent data lineage, decentralized identity management, and automatic execution of contractual agreements using smart contracts [3].

However, integration of blockchain into AIVs is not a silver bullet. It brings its own share of challenges (performance: latency, throughput, privacy) and important architectural questions about how distributed ledgers should behave in the highly dynamic, resource-constrained vehicular environment [4, 5]. Realizing successful implementation is not achievable by a universal definition but by a specialized application of the DLT principles on the less restrictive limitations in the case of robust data demands in autonomous transportation.

However, the application of blockchain technology to AIVs is never a panacea. AIV technology introduces a set of challenges (performance: latency, throughput, privacy) and architectural issues related to how distributed ledgers must act in a highly dynamic and resource-constrained vehicular environment [4, 5]. Successful implementation is actually not achievable by a general definition, but by a specific application of the DLT concepts on the more tolerant boundaries in robust data needs in autonomous transportation.

The rest of the paper is organized as follows. In Sect. 2, we present the background of the paper. Next, the fundamentals of blockchain technology are described in Sect. 3. Section 4 presents the distributed ledger architecture in vehicular networks. While, Sect. 5 provides the security implications of AIV networks. The challenges of the integration of blockchain technology into AIV systems are reported in Sect. 6. Section 7 offers suggestions for future work. Finally, the paper concludes with Sect. 8.

2 Background

Before delving into blockchain architectures and security frameworks for autonomous intelligent vehicles (AIVs), it is important to establish the foundational context. This section highlights the essential concepts that underpin our discussion. First, we review the core principles of blockchain technology, focusing on its decentralized ledger structure, cryptographic integrity, and consensus mechanisms. We then move to the domain of AIVs, outlining how sensors, AI-driven decision making, and vehicle-to-everything (V2X) connectivity enable autonomy while simultaneously creating new trust and security challenges. By situating blockchain within the operational realities of AIV networks, this background provides the technical baseline for the subsequent analysis of integration strategies, architectural considerations, and open challenges.

2.1 Blockchain Fundamentals

Blockchain is a distributed ledger maintained by consensus of a network of nodes without a central authority. Each node has a copy of the ledger, and updates are only confirmed when a consensus is obtained. Cryptography ensures immutability: if someone tries to alter historical records, the hashes will not match, exposing tampering. With this decentralization of trust, reliance on a central server is eliminated. This trust model has already extended beyond cryptocurrencies, finding adoption in supply chain management, healthcare, and intelligent transportation systems [6].

2.2 Autonomous Intelligent Vehicles (AIVs)

Artificial intelligence in intelligent transport systems empowers autonomous intelligent vehicles (AIVs) to sense their environment and make decisions with minimal human intervention. V2X communication encompasses vehicle-to-vehicle (V2V), vehicle-to-infrastructure (V2I), and other modalities that enable collaborative mobility. These capabilities make AIVs attractive for safety-critical applications (e.g., collision avoidance, vehicle platooning) and efficiency improvements (e.g., intelligent traffic management). However, they also introduce vulnerabilities: cyberattacks, data tampering, and privacy threats. Traditional centralized trust schemes such as public key infrastructure (PKI) suffer from single points of failure, long lead times for certificate revocation, and susceptibility to denial-of-service or spoofing attacks [7]. These challenges motivate the exploration of blockchain as an alternative to centralized platforms to ensure security, identification, and trust in AIV ecosystems.

2.3 Blockchain-Enabled AIV Architecture

Building on the key elements of blockchain and AIV, we introduce a unified architecture, which takes advantage of blockchain capabilities for secure and transparent vehicular communication. The architecture of this framework is provided in Fig. 1, which guarantees reliable communication between vehicle, infrastructure, and management authorities.

Each intelligent vehicle (e.g., IV-1, IV-2) generates transactions related to its unique identity. These are validated and stored on the blockchain, providing tamper-

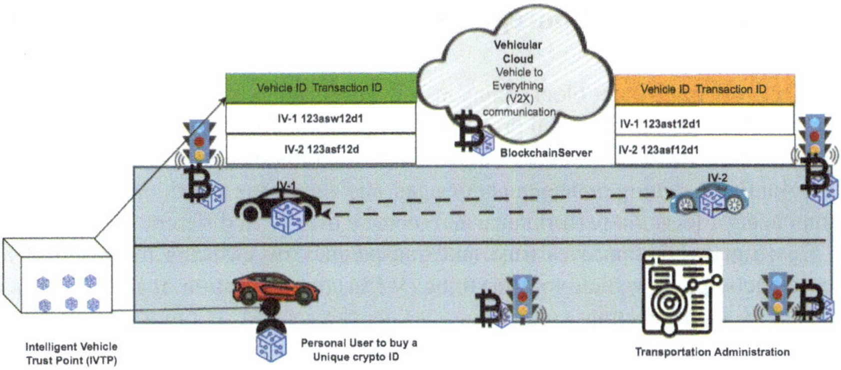

Fig. 1 Blockchain-based architecture for autonomous intelligent vehicles (AIVs)

proof records of vehicular activity. The vehicular cloud functions as the data sharing and computational layer, aggregating communication while offloading resource-intensive tasks. The blockchain server serves as the consensus layer, ensuring the immutability and authenticity of transactions between all participants. The transportation administration, working with unique cryptographic identities, provides regulatory oversight and ensures compliance. Finally, the Intelligent Vehicle Trust Point (IVTP) introduces a reputation-based mechanism to detect and mitigate malicious or compromised nodes.

This architecture improves trust and transparency by guaranteeing transaction integrity in vehicle-to-everything (V2X) communication. Strengthen security against insider threats and spoofing attacks by taking advantage of the immutability of the blockchain. Furthermore, decentralization reduces the dependency on a single centralized authority, distributing trust across the ecosystem. However, practical challenges remain, including addressing latency, throughput, and scalability to support real-time AIV operations.

In summary, the background notes that although decentralization is provided by blockchain, immutability, and trust, its direct use in AIV environments is challenging due to the special nature of vehicular networks, such as high mobility, dynamic connectors, and huge data flow. These challenges require careful design and deployment modifications.

Based on this core blockchain concept, Sect. 3 presents the fundamental blockchain concepts that better support our investigation of AIV systems; including distributed ledger technology, cryptographic building blocks, identity management and consensus devices, as they provide a technical basis for the architectural and security discussions that follow.

3 Key Blockchain Concepts for AIVs

To effectively implement blockchains in AIV networks, it is imperative to comprehend the fundamental technical principles of these systems. The blockchain itself is not a single technology, but rather a combination of distributed ledgers, cryptography primitives, identity management designs, and consensus algorithms. Each of the components affects the performance and security of AIV in different ways.

This framework improves trust and transparency by ensuring the integrity of the transaction over vehicle-to-everything (V2X) communication. It also provides a mechanism for protecting itself against insider threats and spoofing attacks, given the immutability of the ledger.

Moreover, since it is decentralized, trust is less dependent on a single centralized entity spread across the system. There are, however, several practical issues which must be resolved with regard to the problem of being able to satisfy the requirements of latency throughput and scalability in real time for AIV operations.

To address these trade-offs more clearly, in the following subsections we discuss some of the fundamental concepts of blockchain systems with respect to AIVs-by first covering Distributed Ledger Technology (DLT) and Immutability, then presenting cryptographic primitives, identity management, and consensus mechanisms.

3.1 Distributed Ledger Technology and Immutability

Distributed ledger systems such as blockchain allow a network of nodes to maintain a common ledger of transactions without the need for a central authority. Every block is cryptographically connected to the previous, making it tamper-proof and immutable [6, 8, 9].

Formally, a block B_i is defined as:

$$B_i = \left(h_{i-1}, T_i, \mathcal{M}_i\right), \tag{1}$$

where $h_{i-1} = H(B_{i-1})$ is the hash of the previous block, T_i is the set of transactions, and $\mathcal{M}_i$ denotes the block metadata (e.g. timestamp, nonce).

The immutability condition can be expressed as

$$h_i = H(B_i) \quad \Rightarrow \quad h_i \text{ depends transitively on all prior } B_j, \;\; j < i. \tag{2}$$

Thus, any modification to a past block B_j invalidates all subsequent hashes $h_k, k \geq j$, making computationally infeasible tampering.

Proof of travel. Once a trip is confirmed, it cannot be falsified. For example, let d denote vehicle sensor data and t the timestamp; the blockchain stores

$$h_d = H(d \parallel t), \tag{3}$$

which provides irrefutable evidence of "travel" since altering d produces a different hash h_d.

For AIVs, this implies that critical information, such as accident telemetry, can be securely logged and time-stamped, providing stakeholders with a verifiable reference of truth for liability and investigation [10]. It is infeasible to forge or alter such proofs, increasing accountability and transparency.

However, vehicular networks create huge and diverse data streams that cannot be completely stored on-chain. A typical design is to store references to important data or integrity hashes, and keep bulk data off-chain [11]. Formally:

$$h_D = H(D), \quad \text{store } h_D \text{ on-chain, while } D \text{ remains off-chain.} \tag{4}$$

Permissionless distributed ledgers such as Bitcoin and Ethereum have high latency and throughput constraints that are not enough to support real-time vehicular networks. Permissioned and consortium blockchains, on the other hand, only allow participation of legitimate entities (such as manufacturers, regulators, and infrastructure providers), providing faster consensus while preserving decentralization with governance mechanisms [12, 13].

To meet performance requirements, blockchain is typically combined with edge computing. Blockchain nodes may be implemented for lower power devices, via one or more RSUs (roadside units) or cloudlets, to minimize delay and maintain persistence while disconnected. Multilayered architectures and designs—in which local edge blockchains achieve fast logging while global ledgers ensure integrity and consistency—were introduced to achieve a compromise between scalability, integrity, and real-time requirements [14, 15].

3.2 *Cryptographic Primitives and Identity*

The security of blockchains relies on proven cryptographic primitives, such as public key cryptography (usually ECDSA) and the signature mechanism, which are fundamental to trust [6]. In reality, every vehicle or organization possesses a pair of cryptographic keys–a private key sk, usually stored in hardware modules, and a public key pk, which represents its identity in the blockchain [11].

Formally, a key pair is generated as:

$$(pk, sk) \leftarrow \text{KeyGen}(1^{\lambda}), \tag{5}$$

where λ is the security parameter.

Digital signatures can be used both to authenticate and (with certified signatures) for non-repudiation. A digital signature scheme consists of three algorithms:

$$\sigma \leftarrow \text{Sign}(sk, m), \tag{6}$$

$$\{0, 1\} \leftarrow \text{Verify}(pk, m, \sigma), \tag{7}$$

where m is a message, σ is the signature, and Verify outputs 1 if the signature is valid and 0 otherwise.

As an illustrative example, in the context of vehicle-to-anything transmission of a message such as "Obstacle detected at location X," digital signatures provide the required guarantees of message integrity and attribution to a specific cryptographic identity [12]. Message integrity can be formally captured as:

$$\text{Verify}(pk, m, \sigma) = 1 \quad \Rightarrow \quad m \text{ has not been altered.} \tag{8}$$

Compared to PKI-based systems, blockchain identities are typically self-managed and validated by consensus itself to a point where authority-based intermediaries are not employed. In permissioned blockchain systems, a one-time establishment authority is generally employed to admit trusted users (e.g., vehicular networks) [14].

3.3 *Identity Management in AIV Networks*

Identity management is also a serious challenge in an AIV network, as vehicles should balance privacy and accountability well. Static identifiers have challenges including long-term tracking attacks, which is why pseudonymous strategies are preferred and can be supported with, e.g., decentralized identity (DID) systems, zero-knowledge proofs, group signatures [11, 12].

One practical approach is the generation of rotating public keys, minutely timestamped (the timestamp to the second) to the transaction, and recorded by trusted parties on a permissioned blockchain every minute. The vehicles change the keys on the V2X to avoid linkable communications, and the ledger guarantees that the binding is verifiable for the dispute resolution [6, 14].

More sophisticated primitives, such as ring signatures, increase anonymity even more. Automobiles that sense a hazard can sign alerts together, for example, with the blockchain verifying their authenticity without exposing their identity in detail. Threshold ring schemes Ahmed et al. [17], support collaborative authentication and maintain privacy [14, 16, 17]. These two positive properties of anonymity and traceability can be combined in a token: a misbehaving Leave vehicle can still be deanonymized by authorized parties.

The blockchain also works as a decentralized key management system [12]. On-chain identities can be linked to physical vehicle keys or blockchain addresses, and smart contracts can be used to conditionally grant permissions/roles. Compromised keys can be revoked by broadcasting transactions that are visible to all nodes, eliminating delays in traditional PKI systems [6]. This offers a robust, one-way consistent view of legitimate credentials in the very dynamic vehicle world.

3.4 Consensus Mechanisms

Consensus mechanisms are the backbone of distributed ledger systems, enabling nodes to agree on a single version of the truth. In public blockchains, *Proof-of-Work* (PoW) and *Proof-of-Stake* (PoS) are common, but they are generally not suitable for AIVs. PoW requires high computational resources and induces long confirmation delays, while PoS relies on stake-based voting that does not directly capture real-time vehicular trustworthiness [18–20].

The expected confirmation latency of a PoW blockchain can be modeled as follows:

$$L_{PoW} \approx \frac{B}{H \cdot r} + T_{prop}, \tag{9}$$

where B is the block difficulty, H is the total network hash rate, r is the average block generation rate, and T_{prop} is the network propagation delay. This relationship shows why PoW systems often exhibit high latency and low throughput, contradicting the stringent real-time requirements of vehicular environments.

Permissioned Consensus. In AIV permissioned blockchain deployments, *Byzantine Fault Tolerant* (BFT) protocols such as PBFT or Raft are attractive because they provide deterministic finality within seconds. The communication complexity of PBFT can be expressed as follows.

$$C_{PBFT}(n) = O(n^2), \tag{10}$$

where n is the number of participating validator nodes. Although BFT guarantees safety and liveness under Byzantine faults, scalability is limited since communication overhead grows quadratically with network size.

Reputation-Based Consensus. In AIV, reputation systems are integrated into consensus to mitigate Sybil and insider attacks. In *Proof-of-Reputation* (PoR), a node's voting weight is proportional to its reputation:

$$W_i = \frac{R_i}{\sum_{j=1}^{N} R_j}, \tag{11}$$

where R_i is the reputation score of node i and N is the total number of participants. A variant PoR protocol such as MPoR and R-PBFT further enhances resilience by integrating cryptographic hardness assumptions (e.g. SIS) and trust evaluation into BFT [21].

DAG-Based Consensus. Directed Acyclic Graph (DAG) structures, such as the IOTA Tangle, enable parallel transaction confirmation. The expected throughput can be approximated as

$$Th_{DAG} \approx \frac{\lambda \cdot k}{T_{conf}}, \tag{12}$$

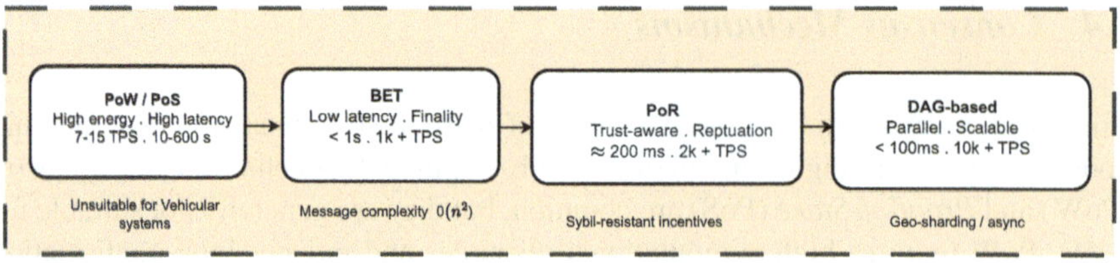

Fig. 2 Evolution of consensus mechanisms for AIVs: from PoW/PoS (unsuitable for vehicular systems) to BFT (low latency), PoR (trust-aware), and DAG-based approaches (scalable), including aspects such as Sybil-resistance and geo-sharding considerations

where λ is the transaction arrival rate, k is the number of parallel validations performed by each new transaction, and T_{conf} is the average time to achieve probabilistic confirmation.

The comparative performance and suitability of these consensus mechanisms for AIV scenarios is summarized in Fig. 2, which contrasts their energy consumption, latency, throughput, and resilience characteristics.

In general, the choice of consensus mechanism in AIV networks is context-dependent. A small and trusted platoon of vehicles may rely on leader-based BFT for low latency, whereas nationwide intelligent transportation systems may require consortium or DAG-based approaches to balance scalability, resilience, and trust. Importantly, while consensus protocols address coordination and performance, they also introduce new security concerns related to attack resistance and trust management. These challenges are examined in detail in the following section (Sect. 5). To anchor the discussion in deployability, Table 1 compares popular consensus algorithms on AIV-related metrics (latency, energy consumption, scalability, and appropriateness). We focus on practical rather than theoretical course trade-offs.

3.5 Smart Contract Applications in AIVs

Smart contracts bring the blockchain concept beyond the secure record keeping to autonomous service composition in AIV. Decentralized logic execution: vehicles, infrastructure, and third-party services can communicate without intermediaries and have a verifiable audit trail.

(i) **Automated Tolling**: AIVs can automatically make a micropayment when they enter a geofenced zone that operates as a toll station, under event stimuli from the chain contract, and without the need for manual intervention delay [22, 23].
(ii) **Usage-based Insurance**: Also known as Pay-per-Use insurance, it enables fairer risk-based pricing, fine-grained rating of premiums, and allowances for vehicle telemetry (speed, location, driver behavior) off the ledger in near real time.

Table 1 Consensus mechanisms compared for AIV deployments (qualitative, deployment-oriented view)

Mechanism	Finality type	Latency (typ.)	Throughput (typ.)	Energy profile	Fault tolerance/assumptions	AIV suitability (why)
Proof of work (PoW)	Probabilistic	High (sec–min)	Low–Med	Very high	Probabilistic safety; Sybil resistance via computation	*Low*. Latency/energy unsuitable for real-time AIV; good as public anchor (periodic notarization)
Proof of stake (PoS)	Often probabilistic[1]	Med (sec)	Med–High	Low–Med	Sybil resistance via stake; sensitive to stake concentration	*Medium*. Better latency/energy; still not deterministic finality in many networks
Delegated PoS (DPoS)	Prob./quasi-final	Low–Med (sub-sec–sec)	High	Low	Small validator set; governance dependence	*Medium–High*. Good performance; governance centralization risk for safety-critical ops
PBFT/BFT variants (e.g., Tendermint)	Deterministic	Low (10–300 ms)[2]	High (k–10k tps)[2]	Low	Tolerates $f < n/3$ Byzantine; known validators	*High* in RSU/MEC domains. Deterministic finality fits control loops and liability trails
HotStuff (BFT lineage)	Deterministic	Low (tens–hundreds ms)	High	Low	Linear messaging; $f < n/3$	*High*. Scales validator count more cleanly; good for regional/consortium AIV
Raft (Crash-fault tolerant)	Deterministic	Very low (ms)	Very high	Low	CFT only; leader-based	*High* for *intra-operator* logs (non-Byzantine). Not suited for adversarial V2X
Proof of authority (PoA)	Deterministic	Low (sub-sec)	High	Low	Trusted authorities; identity-bound validators	*High* in regulated corridors (tolling/ports), where operators are pre-vetted
DAG/tangle (aBFT)	Eventual/fast	Low–Med	High	Low	Gossip + tips selection; aBFT assumptions	*Medium*. Good for high-rate telemetry; careful tuning needed for fast-moving nodes
Proof of elapsed time (PoET)	Prob./scheduled	Low–Med	Med–High	Low	Trusted execution (TEE) reliance	*Medium*. Performance OK; depends on TEE supply-chain trust

Notes Values are indicative and deployment-dependent (validator count, network RTT, message size).
[1] Some PoS networks provide rapid finality layers; others require multiple confirmations.
[2] Typical for regional validator sets (~4–50 nodes) on LAN/MEC backhauls

(iii) **Accident Liability Assignment**: Sensor data and event data recorders (EDRs) stored on the blockchain are irrefutable evidence that allows smart contracts to automatically assign liability to all stakeholders [24].
(iv) **Energy and Charge**: In the case of an EV environment, AIVs can purchase or sell energy to/from each other on peer-to-peer (P2P) charging markets as tokens represented by credits whose exchanges are operationally governed by smart contracts and vehicle identifications [25].
(v) **Fleet Leasing and Regulatory Compliance**: Firms that own a vehicle fleet can use zero-knowledge proof (ZKP) to automate lease settlements or proof of compliance (e.g., inspection results, emission reports) [26].

These included applications that apply to AIV and highlight how decentralized automation represents a natural partner to vehicle intelligence that improves transparency, reduces administrative friction, and provides accountable and auditable operations for large-scale mobility systems.

4 Distributed Ledger Architectures for Vehicular Networks

With the basis laid in the previous two sections about how a consensus-making mechanism can look at and interact with an application (3.4) and with the implementation of smart contracts (3.5), we switch from functional logic to architectural deployment at this point. Although mobile and wireless characteristics are commonly encountered in the Internet of Things (IoT) and Mobile Adhoc NETwork (MANET), decentralized ledgers have a profound implication on how we design system interactions within vehicular ecosystems.

Figure 3 presents a high-level view of the three main ledger architectures, *permissionless*, *permissioned*, and *consortium*, and their relationship with vehicular communication types (V2V, V2X and the assisting base station). Each model supports a set of blockchain-based applications, such as accident reports, trust services, traffic information coordination, and secure software updates.

Drawing from the commonly archived consensus view, we present Table 2 to compare public, private, and consortium blockchain models, highlighting governance, performance, privacy, and security use cases for consideration by AIV stakeholders in choosing a given model.

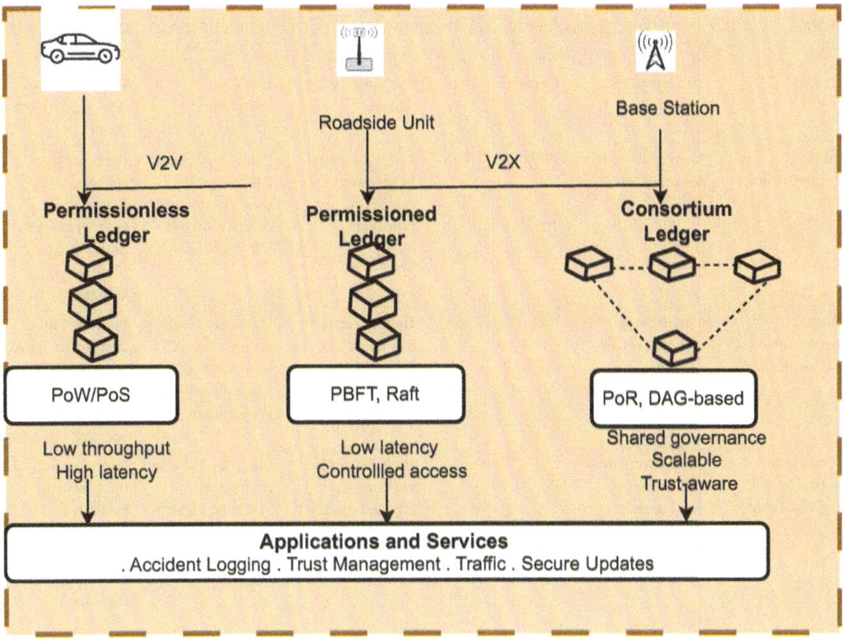

Fig. 3 Distributed ledger architectures for vehicular networks, highlighting *permissionless*, *permissioned*, and *consortium* ledger models. Each architecture corresponds to distinct vehicular communication modes (V2V, V2X, and base station–assisted) and connects to shared applications and services, including accident logging, trust management, traffic coordination, and secure updates

4.1 Permissionless Ledger

Permissionless blockchains, such as those based on PoW or PoS, allow any participant in the AIV network to join and authenticate transactions without prior authorization. In the AIV context, this architecture aligns naturally with vehicle-to-vehicle (V2V) communications, where trust must be established among a large number of potentially unknown entities.

The consensus latency L_{pl} for a permissionless blockchain can be approximated as:

$$L_{pl} \approx \frac{D}{B} + T_{prop}, \tag{13}$$

where D is the block size, B is the average block generation rate, and T_{prop} is the network propagation delay. As highlighted in Fig. 3, permissionless ledgers support use cases such as public accountability for traffic incidents, although their high latency and limited throughput present challenges for real-time safety applications.

Table 2 Ledger deployment models in AIVs: governance, performance, privacy, and security use cases

Model	Governance/access	Consensus (typ.)	Performance/ latency	Privacy/ compliance	Representative AIV security use cases
Public (permissionless)	Open participation; token incentives	PoW, PoS, DPoS	Low throughput; high latency (global propagation)	Public by default; ZK/rollups for selective privacy; complex GDPR alignment	Public audit anchors, OTA notarization, cross-ecosystem proofs (e.g., timestamping EDR hashes)
Private (permissioned, single operator)	Single entity (OEM, fleet, or city DOT)	Raft, PBFT, PoA	High throughput; low latency in LAN/MEC	Granular access control; data minimization; strong policy enforcement	Intra-fleet logs, compliance trails, low-latency tolling/charging within one operator's domain
Consortium (permissioned, Multi-stakeholder)	Joint governance (OEMs, insurers, RSU operators)	PBFT, HotStuff, PoA	High throughput; sub-second regional finality	Controlled sharing; contractual visibility; clear liability frameworks	Inter-operator V2X trust, incident evidence exchange, pay-per-use insurance, cross-border tolling

Guidance: *Private* ledgers suit single-operator, ultra-low-latency control loops; *consortium* ledgers enable accountable multi-party coordination; *public* ledgers act as audit anchors or cross-ecosystem integrity proofs

4.2 *Permissioned Ledger*

Permissioned blockchains restrict participation to authenticated entities, such as automotive manufacturers, regulatory bodies, and certified roadside units. Consensus mechanisms such as Practical Byzantine Fault Tolerance (PBFT) or Raft enable faster transaction confirmation with lower latency compared to public networks.

The throughput Th_{pm} of a permissioned ledger is illustrated as:

$$Th_{pm} = \frac{N}{L_{pm} + \Delta}, \tag{14}$$

where N denotes the number of transactions per block, L_{pm} is the average consensus latency, and Δ is the communication delay between authorized participants. As indicated in Fig. 3, permissioned ledgers are suitable for vehicle-to-everything (V2X) interactions that require reliable, low-latency coordination.

4.3 Consortium Ledger

Consortium blockchains occupy an intermediate position between fully public and fully private models. They are governed by a consortium of pre-selected organizations, such as automotive companies, telecom operators, and government agencies. Consensus protocols such as Proof-of-Reputation (PoR) or Directed Acyclic Graph (DAG)-based approaches enable high scalability and trust-aware collaboration across multiple domains.

A simplified trust score $T_c(i)$ for a consortium participant i can be defined as:

$$T_c(i) = \frac{\alpha \cdot R_i + \beta \cdot H_i}{\alpha + \beta}, \tag{15}$$

where R_i is the participant's reputation score i, H_i is the historical honesty metric, and α, β are tunable weighting factors. This formulation emphasizes the role of governance and accountability in ensuring security and fairness within the consortium.

As highlighted in Fig. 3, the consortium ledger is best suited for base station-assisted communication and cross-organizational cooperation, supporting scalable functions such as secure data sharing, collaborative traffic management and large-scale software updates.

In summary, consortium blockchains merge the transparency of public ledgers with the robustness of permissioned architectures. But their practice in AIV networks also depends on how the underlying infrastructure can support communications and consensus among interconnected parties.

4.4 Infrastructure and Network Topology

Following the discussion of ledger types above, this section investigates how infrastructure and network topology can facilitate the deployment of blockchain in intelligent vehicular systems. In contrast, the blockchain in an AIV environment needs to support highly dynamic nodes, such as moving vehicles with VANET radios, e.g., DSRC or C-V2X, as well as fixed roadside and cloud infrastructure. One possible solution is a *hybrid topology* in which stationary proxy nodes (e.g. RSUs or edge servers) maintain the blockchain backbone, while vehicles are position as lightweight clients with intermittent operation [27, 28].

For example, consider a city that is deployed with RSUs every few kilometers. These RSU's can be organized a mesh network that is connected to a blockchain consensus, it can be run all the time. As they drive in traffic, they concur with the nearby RSU by sending new transactions (e.g., (Car A at location L at time T)) and getting informed about changes in the local ledger state. Thus, vehicles have a localized but trusted perspective of the blockchain, even if they have transient disconnections. The RSUs themselves are connected to each other via high-speed fiber or 5G connections, resulting in fast block propagation and without significant latency that would be found in vehicle-to-vehicle multi-hop communication [27].

In resource-limited infrastructure areas, vehicles can create ad hoc blockchain clusters and keep local micro-ledgers. They are low-overhead consensus groups that allow vehicles in close proximity to share state and make transactions without needing to have global connectivity. The micro-ledger can commit summaries or checkpoints to the global blockchain whenever it connects. This kind of *zonal sharding* scales better by confining the consensus only in regional data [29].

Cross-zone coordination should possess an interoperability setup to secure reputation, identity, and transaction validation across timestamps of the same vehicle between different blockchain shards. The cross-shard protocols and dynamic graph partitioning bring trusted state transfer across the regional ledgers, so the authenticated reputation of a vehicle in one zone is valid in another [29].

In summary, AIV systems for blockchain deployment would benefit from applying a hierarchical and hybrid architecture. Static RSUs and edge servers are the infrastructure part to establish a secure consensus, while vehicles are the dynamic and lightweight participants. Micro-ledgers controlled by local jurisdictions enhance scalability, and protocols between shards achieve cross-domain trust and mobility. These architectural modifications are capable of handling the distinctive features of the blockchain with respect to mobility, scalability, and trust in vehicular networks.

4.5 Data Management and Off-Chain Storage

Extending the architectural discussion, this subsection examines how AIV systems manage the vast amounts of sensor data they continuously generate. In conventional blockchain networks (e.g., cryptocurrencies), the nodes are relatively static and maintain well-connected relations via the Internet. In contrast, AIV environments are characterized by high mobility and heterogeneous data sources—including LiDAR, cameras, and V2V beacons—which makes full on-chain storage impractical due to throughput and block size constraints.

To address these challenges, most platforms adopt a *on-chain/off-chain separation strategy*. Here, only essential information such as cryptographic hashes or aggregated results is recorded on-chain, while the bulk of raw data (e.g., high-definition video streams or sensor telemetry) is stored in decentralized databases or cloud infrastructure [30]. This separation can be formally expressed as follows.

$$D_{\text{total}} = D_{\text{on-chain}} + D_{\text{off-chain}}, \tag{16}$$

where D_{total} represents all vehicle data generated, $D_{\text{on-chain}}$ is the critical subset stored immutably on the blockchain, and $D_{\text{off-chain}}$ denotes the larger bulk non-critical data maintained externally.

In addition to separation, *zonal sharding* has emerged as a complementary technique for managing large-scale vehicular data. In this approach, vehicles within a given geographical zone form lightweight microledgers that handle local transactions and periodically commit checkpoint summaries to the global blockchain. The communication cost of intra-zone BFT consensus with n total nodes partitioned into Z zones can be approximated as:

$$C_{\text{intra}}(n, Z) = O\left(\frac{n^2}{Z}\right), \tag{17}$$

where each zone contains $\frac{n}{Z}$ nodes on average. This reduces quadratic messaging overhead by a factor of Z, while cross-zone coordination ensures interoperability and preserves vehicle reputation and transaction validity across different shards [29].

Furthermore, certain high-frequency control messages (e.g., safety beacons, cooperative lane changes, or intersection coordination) require even lower latency than block-level confirmation. Such data flows can be supported using *state channels*, where parties exchange numerous off-chain transactions and only commit final aggregated results to the blockchain, thus alleviating on-chain congestion [31].

Finally, ensuring data privacy and cross-chain compatibility remains essential. Sensitive information, such as vehicle location histories, must be shared selectively with authorized parties using privacy-preserving cryptographic techniques [32]. Likewise, as AIVs move across administrative boundaries, cross-chain protocols and governance frameworks are required to maintain consistent reputation, identity, and transaction validity across diverse ledger systems [33, 34].

In summary, combining on-chain/off-chain separation with zonal sharding and state-channel designs provides a scalable and efficient pathway for handling high-volume, high-frequency data from vehicular networks. However, these architectural strategies inevitably introduce new attack surfaces and inter-zone vulnerabilities, making it crucial to examine the broader security implications of blockchain-enabled AIV systems, as discussed in the following section.

5 Security Implications for AIV Networks

Building on the data management and architectural considerations discussed in the previous subsection, this section focuses on one of the most critical dimensions of blockchain integration in AIVs: *security*. The primary value of distributed ledgers lies in their ability to provide transparency, accountability, and trust among heterogeneous participants. In particular, the immutability of blockchain ensures that once critical data, such as event logs, sensor readings, or control commands, are recorded, they cannot be altered without detection, making large-scale attacks extremely difficult in well-functioning vehicular environments [35, 36].

5.1 Accountability

A key security benefit is accountability. Each blockchain transaction is bound to a cryptographic identity through digital signatures, thereby linking every vehicular action to a verifiable entity. For example, in a cooperative driving scenario where vehicles agree on merge priorities, any malicious deviation leading to a collision would be permanently recorded on the ledger. The responsible entity cannot repudiate its actions, and investigators obtain a transparent and tamper-proof audit trail [37].

5.2 Resilience

Another dimension is resilience. Traditional centralized approaches create single points of failure: a compromised server or controller can disrupt the entire traffic management system. In contrast, blockchain relies on group consensus, where the system remains operational even when some nodes act maliciously or fail. This distributed robustness is vital for maintaining the availability of safety-critical AIV operations [38].

5.3 Trust and Privacy

Beyond resilience, blockchain improves trust in data exchange. Vehicles can directly verify safety messages (e.g. traffic alerts or intersection status) against immutable ledger entries rather than relying solely on centralized authorities. Reputation mechanisms strengthen this property: the influence of each node W_i on consensus can be modeled as

$$W_i = \frac{R_i}{\sum_{j=1}^{N} R_j}, \tag{18}$$

where R_i is the reputation score of node i and N is the total number of participants [?] [39]. Such mechanisms discourage malicious behavior and mitigate Sybil attacks [40]. At the same time, blockchain transparency raises privacy concerns: Without adequate safeguards, sensitive data, such as vehicle trajectories, can be exposed. Privacy-preserving tools, including encryption, selective disclosure, and ZKPs, are therefore essential [41].

5.4 *Identity and Credential Management*

A further challenge lies in managing vehicular identities. Blockchain can function as a decentralized certificate authority, storing immutable vehicle attributes, from inspection history to autonomy level, on a permissioned ledger [42]. Keys can be securely distributed, rotated, or revoked through blockchain transactions, allowing fine-grained control over access rights and ensuring system-wide trust [43].

In summary, blockchain provides AIV networks with tamper-proof records, verifiable accountability, resilience, and decentralized trust. However, these advantages are accompanied by new risks, particularly concerning smart contract vulnerabilities, key compromise, and privacy leakage. Addressing these issues requires careful design of security frameworks, which is the focus of the following section on Challenges in Integrating Blockchain with AIVs.

5.5 *Security and Regulatory Considerations*

The deployment of blockchain technology in an AIV system introduces new perspectives and challenges for end-to-end vehicular security. Unlike static IoT networks, AIVs function in a dynamic topology-constantly moving-mobility, and heterogeneous interfaces (V2V, V2I, and V2X) being full examples, such that decision making for safety measures must respond aggressively to ensure low-latency. As a result, security must cover not only the classical properties of the CIA (confidentiality, integrity, and availability), but also trust, verifiability, and resistance against Byzantine and real-time attacks [44]. The broader description of the entire concept of security and regulatory consideration of blockchain technology in AIV is further detailed below.

(i) **Dynamic Topology and Consensus Resilience**. Consensus algorithms must be designed with respect to vehicular mobility and sporadic connectivity. Traditional blockchain protocols work with relatively steady sets of validators, whereas in AIV networks, the participants (vehicles coming in and out coverage zones) are highly dynamic. Directed Acyclic Graph (DAG)-based ledgers, such as IOTA Tangle and the Avalanche system subnets, provide higher throughput and aBFT asynchronously for high mobility. However, they experience difficulties in guaranteeing transaction finality in the presence of fast topology changes and a possible desynchronization of fast wandering users. To address such challenges, techniques such as optimized gossip protocols [45], dynamic tip selection algorithms, and localized validator clustering have been introduced; however, they have not yet been explored on the highway scale.

(ii) **Scalability in Dense Urban Environments** Byzantine Fault-Tolerant (BFT) consensus algorithms such as PBFT, HotStuff, and Tendermint can offer deterministic finality and can be particularly appropriate for the localized deployments of AIVs (e.g., RSU-based or MEC-orchestrated networks). However, they are not efficient when applied to city-scale scale topologies due to quadratic message complexity [46]. To maintain sub-second confirmation times while retaining accountability, hierarchical BFT (HBFT) and sharded consensus have been considered [47]. In the case of AIV, this trade-off directly affects the extent to which inter-vehicle trust exchange and coordinated intersections and congested traffic decisions can be trusted [5, 10].

(iii) **Privacy, GDPR, and Data Protection** Vehicular networking is based on sensors that contain sensitive and personal data, such as driver identity, vehicle telemetry, and location traces. Compliance with the EU General Data Protection Regulation (GDPR) and the UN Regulation No. 155-Cybersecurity and Cyber Security Management System for Vehicles of the United Nations Economic Commission for Europe regulations is crucial [48]. Decentralized ledgers do not show potential to manipulate, but, relatedly, they are not easily expungeable as well, in direct contradiction to the 'right to erasure' according to GDPR. Hence, privacy-preserving schemes such as zero-knowledge proofs (ZKPs), ring signatures, and selective disclosure systems are incorporated to comply with data sharing [49]. Off-chain storage using on-chain hashes also meets the data minimization requirement with maintaining integrity as well.

(iv) **Safety and Automotive Regulatory Standards** In the area of safety, blockchain-based cooperation in AVs must be compatible with established automotive safety standards, including ISO26262 (functional safety), ISO/SAE21434 (road vehicle cybersecurity) and ISO 21448 (intended functionality safety, SOTIF) [4]. These guidelines include formal verification of safety-critical subsystems and traceability for all software modifications to communication records. These needs are supported using distributed ledgers, which can offer immutable and time-stamped evidence of, e.g. over-the-air (OTA) updates going through some part of the supply chain, liability attribution for incidents, and also post-incident forensics [50]. Furthermore, integration with runtime assurance frameworks guarantees the certifiability of blockchain-based decision layers in the automotive safety life-cycle.

(v) **Key Insights** In conclusion, AIV-specific security requires some multilayered design considerations: (i) mobility-aware consensus methods such as DAG and HBFT maintain throughput in the dynamic topologies; (ii) privacy-preserving designs to bridge blockchain immutability with GDPR and automotive compliance; and (iii) safety alignment with ISO/SAE standards that would allow proving and certifying vehicular autonomy. These guidelines highlight that blockchain in AIVs has the potential not only as a data integrity tool, but primarily as an enabler of secure, auditable, and regulatory-compliant autonomous mobility [5].

6 Challenges in Integrating Blockchain with AIVs

Despite its potential, the incorporation of blockchain in AIV systems is not as straightforward. Both phenomena collide with an intersection of technical, security, and regulatory challenges that must be solved before blockchains can be used as a robust basis for safety-critical applications in mobility [36]. Some of the major challenges are the following.

(i) **Scalability and Throughput**: In a large-scale AIV environment, millions of vehicles can generate transactions every second, ranging from sensor updates to service payments. However, current public blockchains sustain only 10^1 -10^2 transactions per second, orders of magnitude below vehicular requirements. Even consortium blockchains exhibit loss of performance as the network size increases [51, 52]. Research on sharding, hierarchical architectures, DAG-based ledgers, and layer-2 protocols seeks to close this gap and enable blockchains to operate at the vehicular scale [53].
In a large-scale AIV system, there could be millions of vehicles that generate transactions every second, ranging from sensor updates to service payments. Unfortunately, existing public blockchains best support 10^1 -10^2 transactions per second, which is several orders of magnitude lower than the requirements for vehicular scenarios. Even the workload of consortium blockchains reduces as the network scales [51, 52]. The existing literature on sharding, hierarchical architectures, DAG-based ledgers and layer-2 protocols aims to bridge this gap and transform the work of blockchains to the vehicular scale [53] [?].

(ii) **Latency and Real-Time Constraints**: Safety critical applications like collision avoidance and intersection control require sub-second reaction times. Traditional confirmation delays, which may last for a minute, are not acceptable in this scenario. Fast-finality protocols (for example, BFT) [46, 54] and hybrid designs, which treat time-critical logic off-chain and record the result afterwards [10], are required to balance the security aspects of the blockchain with the real-time needs of vehicular networks.

(iii) **Privacy Concerns**: While the transparency of Blockchain helps to hold the actors accountable, it also runs a risk of revealing sensitive information, such as vehicle movements, driving patterns, and service usage. Pseudonymous identifiers may also be linked to uncover private past histories. To address such a problem, some privacy-preserving approaches, such as key rotation, ring signature, and zero-knowledge proof, have been designed to protect contact information in preserving privacy without compromising trust and verifiability of contacts [41].

(iv) **Security and Attack Resistance**: Although blockchain is inherently immutable, it can still be subject to sophisticated forms of attacks. Sybil and insider attacks, as well as denial-of-service (DoS) attacks, continue to present significant threats, and compromised cryptographic keys may cause a whole node to be compromised to some large extent. Security defense mechanisms can be implemented as admission control, trusted key management hardware, and efficient key revocation facilities for resilience [35, 37, 38, 40, 43, 55, 56].

(v) **Legal and Regulatory Hurdles**: There are unanswered legal and regulatory issues in the adoption of blockchain for transportation. Challenges regarding smart contracts' enforceability, liability allocation, and compliance with cross-jurisdictional frameworks like GDPR have not been addressed. Regulators may also request special features, such as auditability or selective erasure, that are in conflict with the blockchain tenets of immutability and consensus design [53].

In short, blockchain's contribution to the development of AIVs depends not only on technological bottlenecks, but also on the ability to overcome more general privacy and security issues and the development of an effective governance regime. Thus, it is important to address these challenges in order to build a solid, secure, scalable, and legally strong base for future vehicular networks.

7 Future Directions

Looking ahead, several research avenues are poised to shape the role of blockchain in autonomous and intelligent vehicle (AIV) systems.

7.1 Integration with AI and Machine Learning

This intersection of blockchain and AI/ML offers the ability to create adaptive trust management. Anomalous patterns in vehicle data can be identified by machine learning and integrated with reputation systems on the blockchain, or may trigger smart contract responses [37]. Blockchains, on the contrary, can use verifiable data sets for training. For example, federated learning can make use of blockchain to synchronize model updates and grant fair payments between vehicles [53]. Promising ideas involve AI-supported consensus protocols that automatically tailor the selection of a leader and block sizes to current, desires conditions.

7.2 Edge Computing and 6G for Ultra-Low Latency

Beyond 5G (e.g., 6G) networks with ultra-reliable low-latency communication (URLLC) provide opportunities for real-time blockchain in vehicular clusters. Edge nodes, such as RSUs, can execute localized blockchains with millisecond level confirmations, whereas 6G backhaul connections will synchronize the global states. This type of integration could allow the development of safety-critical applications, such as cooperative/collaborative maneuvers or swarm coordination, to log and validate the behavior of these applications in almost real time [57].

7.3 Standardization and Consortium Efforts

Both interoperability and regulatory approval depend on standardization. Efforts such as IEEE P2418.5 and the Mobility Open Blockchain Initiative (MOBI) are establishing standards for digital vehicle identity, data formats, and cross-border use cases [53]. The consortium blockchains, led by car manufacturers, infrastructure providers and insurers, are expected to act as trusted platforms for applications, including usage-based insurance, vehicle history, and accident reporting.

7.4 Privacy-Enhancing Cryptography

The challenge, as always, is to maintain functionality while preserving privacy. New tools and mathematical paradigms such as zero-knowledge proofs (zk-SNARKs, zk-STARKs), secure multiparty computation (MPC), and homomorphic encryption enable such protection and are being developed [58]. This use case would enable vehicles to demonstrate compliance, for example, with safety or emissions constraints, without disclosing sensitive input data, or to calculate aggregate traffic statistics without revealing individual (and potentially privacy-sensitive) contributions. Efficient group and ring signature schemes that may be additionally quantum secure, are also likely to have an expanding importance in AIV blockchains in the future.

In summary, AIV ecosystems with secure and trustworthy capabilities are restricted by the fusion between blockchain and AI/ML, edge-assisted 6G scenarios, standardized architectures and privacy-preserving cryptography. These guidelines seek to address the trade-off between privacy, scalability, and trust, making the blockchain a fundamental piece of an autonomous mobility infrastructure.

8 Conclusion

Blockchain technology provides an attractive solution to some of the trust and security problems in AIV networks. Blockchain can also be adopted to guarantee the trustworthiness aspect of the information and the accountability as well as coordination (in case of vehicles, and infrastructures have not been fully trusted) established by the enabling smart contracts and decentralized tamper evidence ledger. In this extended article, we have explored the technical underpinnings of blockchain (distributed ledger structures, cryptographic primitives, and consensus algorithms) and related them to the requirements of AIV systems. We also discussed how smart contracts could automatically carry out complex interactions, traveling as one, paying tolls, insurance, and sharing data in confidence. The adoption of blockchain technology within the vehicle environment requires carefully crafted designs that rely on edge computing, off-chain storage, and permissioned consensus mechanisms to achieve the desired performance and privacy properties. There were many challenges that we identified, everything from scalability and real-time requirements to privacy, legal, and regulation, that need to be figured out for this to be adopted.

However, despite the challenges, there is promising progress in both research and pilot applications. Blockchain-secured V2X communications, decentralized ride-sharing platforms, and the vehicular data marketplace have been prototyped with cryptocurrency incentives. The lessons are incorporated into emerging best practices around these efforts. A clear trend is the integration of blockchains with other technologies, such as AI and machine learning, to dynamically assess the veracity of data and identify abnormalities for vehicular networks and with advanced wireless and edge computing infrastructure for the required bandwidth and support for real-time processing. Another trend, which is addressed, is an effort toward standardization, something like IEEE P2418.5 (Blockchain in Transportation), and consortiums such as MOBI are developing standards for vehicle identity, data formats, and transaction rules to make sure that different automakers and jurisdictions can operate coherently and consistently.

Further future research directions include the enhancement of privacy protection (e.g., through zk proofs for protecting sensitive vehicular data), scalability of blockchain consensus for large-scale vehicular networks, as well as industry standards and regulatory framework definition for blockchain in transportation. By actively solving the issues mentioned today and by moving forward with pilot solutions in real-world environments, stakeholders can guarantee that blockchain unlocks the potential for more secure, trusted, and self-sustaining autonomous mobility platforms. Still, many questions remain, but the advances we have seen to date imply that blockchain has the potential to be a core enabler of the digital infrastructure fueling the future of driverless vehicles.

References

1. Bellagarda, J.S., Abu-Mahfouz, A.M.: An updated survey on the convergence of distributed ledger technology and artificial intelligence: current state, major challenges and future direction. IEEE Access **10**, 50773–50796 (2022)
2. Alt, R., Gräser, M.: Distributed ledger technology. Electron. Mark. **35**(1), 53 (2025)
3. Singh, P.K., Singh, R., Nandi, S.K., Ghafoor, K.Z., Rawat, D.B., Nandi, S.: Blockchain-based adaptive trust management in internet of vehicles using smart contract. IEEE Trans. Intell. Transp. Syst. **22**(6), 3616–3630 (2020)
4. Singh, A., Sural, S., Sengupta, T., Sural, S.: AVChain: trusted sharing of autonomous vehicle crash incident data using interoperating HyperLedger fabric networks and IPFS. Distrib. Ledger Technol. Res. Pract.

5. Iordache, S., Patilea, C.C., Paduraru, C.: Enhancing autonomous vehicle safety with blockchain technology: securing vehicle communication and ai systems. Futur. Internet **16**(12), 471 (2024)
6. Yaga, D., Mell, P., Roby, N., Scarfone, K.: Blockchain technology overview (2019). arXiv:1906.11078
7. Author, F., Author, L.: Security challenges in vehicular networks: a review of PKI and blockchain approaches. Salford Repository of Research Outputs (2020). Accessed 4 Sept 2025
8. Singh, A.P.: Blockchain technology: core mechanisms, evolution, and future implementation challenges (2025). arXiv:2505.08772
9. Antal, C., et al.: Distributed ledger technology review and decentralized applications development guidelines. Research overview via ResearchGate (2025)
10. Biswas, A., Wang, H.-C.: Autonomous vehicles enabled by the integration of IoT, edge intelligence, 5G, and blockchain. Sensors **23**(4), 1963 (2023)
11. Khan, M.A., Salah, K., Rehman, M.H.U., Jayaraman, R., Omar, M.: A survey on the blockchain techniques for the internet of vehicles. Veh. Commun. **27**, 100285 (2021). https://doi.org/10.1016/j.vehcom.2020.100285
12. Wang, H., Qi, Q., Ma, R., Gao, M., Meng, W., Liu, Y.: A systematic review on blockchain-enabled internet of vehicles (BIoV): challenges, defenses, and future research directions. IEEE Trans. Intell. Transp. Syst. **24**(5), 5073–5095 (2023). https://doi.org/10.1109/TITS.2022.3147890
13. Iordache, M., Popescu, A., Sandu, M., Cioara, T.: Enhancing autonomous vehicle safety with blockchain and edge computing: A decentralized trust framework. Futur. Internet **16**(12), 471 (2024). https://doi.org/10.3390/fi16120471
14. Iordache, M., Reijsbergen, D., Gamal, H.M., Hsu, C., Mu, Y., Ko, R.K.L.: A blockchain-enabled vehicular edge computing framework for secure performance-oriented v2x service delivery. Futur. Internet **14**(12), 471 (2022). https://doi.org/10.3390/fi14120471
15. Surapaneni, P., Bojjagani, S., Bharathi, V., Morampudi, M.K., Maurya, A.K., Khan, M.K.: A systematic review on blockchain-enabled internet of vehicles (BIoV): challenges, defenses and future research directions. IEEE Access (2024)
16. Zhang, X., Lai, J., Moshayedi, A.J.: Traffic data security sharing scheme based on blockchain and traceable ring signature for VANETs. Peer Peer Netw. Appl. **16**, 2349–2366 (2023). https://doi.org/10.1007/s12083-023-01532-7
17. Ahmed, I., Khan, M.I., Lee, S.: Blockchain-integrated threshold ring signatures for secure and anonymous traffic event validation in VANETs. In: 2022 IEEE International Conference on Communications (ICC), pp. 1–6. IEEE (2022). https://doi.org/10.1109/ICC45855.2022.9838962
18. Lin, C., He, D., Huang, X., Zhang, K., Choo, K.-K.R., Vasilakos, A.V.: When internet of things meets blockchain: challenges in distributed consensus. IEEE Netw. **33**(6), 133–139 (2019). https://doi.org/10.1109/MNET.2019.1800246
19. Li, X., Lu, R., Liang, X., Shen, X.S.: Blockchain-based internet of vehicles: Consensus, security and privacy. IEEE Internet Things J. **7**(3), 2694–2703 (2020). https://doi.org/10.1109/JIOT.2019.2946550
20. Ajakwe, S.O., Igboanusi, I.S., Lee, J.-M., Kim, D.-S.: iBANDA: a blockchain-assisted defense system for authentication in drone-based logistics. Drones **9**(8), 590 (2025)
21. Hîrţan, L.-A., Dobre, C., González-Vélez, H.: Blockchain-based reputation for intelligent transportation systems. Sensors **20**(3), 791 (2020). https://doi.org/10.3390/s20030791
22. Graham, D.: Integrated Electronic Payment Technologies for Smart Cities. Springer (2023)
23. Gurumurthy, K.M.: Shared Autonomous Vehicle System Designs for Major Metro Areas: an Examination of Geofencing, Ride-sharing, Stop-location, and Drivetrain Decisions. The University of Texas at Austin (2020)
24. Philip, A.O., Saravanaguru, R.K.: Smart contract based digital evidence management framework over blockchain for vehicle accident investigation in IoV era. J. King Saud Univ. Comput. Inf. Sci. **34**(7), 4031–4046 (2022)
25. Zhang, J., Che, L., Wan, X., Shahidehpour, M.: Distributed hierarchical coordination of networked charging stations based on peer-to-peer trading and EV charging flexibility quantification. IEEE Trans. Power Syst. **37**(4), 2961–2975 (2021)

26. Wali, H., Kulkarni, V., Ganiger, R., Iyer, N.C.: ZKAV: zero knowledge proof for AV. Procedia Comput. Sci. **237**, 891–898 (2024)
27. He, L., Li, F., Xu, H., Xia, W., Zhang, X., Tao, X.: Blockchain-based vehicular edge computing networks: the communication perspective. Sci. China Inf. Sci. **66**(7), 172301 (2023). https://doi.org/10.1007/s11432-022-3658-7
28. Sehar, N., et al.: Blockchain enabled data security in vehicular networks. Sci. Rep. **13**, 4312 (2023). https://doi.org/10.1038/s41598-023-31442-w
29. Han, H., Chen, S., Xu, Z., Dong, X., Zeng, J.: Trust management scheme of iov based on dynamic sharding blockchain. Electronics **13**(6), 1016 (2024). https://doi.org/10.3390/electronics13061016
30. Tram, T., Bellavista, P., Foschini, L., Zanni, A.: State of data platforms for connected vehicles and infrastructures. Comput. Commun. **182**, 34–48 (2021). https://doi.org/10.1016/j.comcom.2021.01.014
31. Dziembowski, S., Faust, S., Hostáková, K.: Perun: virtual payment hubs over cryptocurrencies. In: 2019 IEEE Symposium on Security and Privacy (SP), pp. 106–123. IEEE (2019). https://doi.org/10.1109/SP.2019.00019
32. Dorri, A., Kanhere, S.S., Jurdak, R.: Blockchain for IoT security and privacy: the case study of a smart home. Proc. IEEE **107**(6), 1199–1210 (2019). https://doi.org/10.1109/JPROC.2019.2912227
33. Zheng, Z., Xie, S., Dai, H.-N., Chen, X., Wang, H.: Blockchain interoperability: a survey. IEEE Trans. Emerg. Top. Comput. **9**(3), 1113–1125 (2021). https://doi.org/10.1109/TETC.2020.2967072
34. Belchior, R., Vasconcelos, A., Guerreiro, S., Correia, M.: A survey on blockchain interoperability: past, present, and future trends. ACM Comput. Surv. **54**(8), 1–41 (2021). https://doi.org/10.1145/3471140
35. Xie, J., Tang, H., Huang, T., Yu, F., Dai, H.-N., Xiong, N.: Security and privacy on blockchain-based systems: a survey. IEEE Access **7**, 47615–47631 (2019). https://doi.org/10.1109/ACCESS.2019.2909477
36. Ajakwe, S.O., Kim, D.-S.: Facets of security and safety problems and paradigms for smart aerial mobility and intelligent logistics. IET Intel. Transport Syst. **18**, 2827–2855 (2024)
37. Yu, F., Xie, J., Huang, T., Xiong, N., Liu, X.: Blockchain for vehicular networks: a survey on security, privacy, and applications. IEEE Access **8**, 145822–145843 (2020). https://doi.org/10.1109/ACCESS.2020.3015424
38. Conti, M., Sandeep, K., Lal, C., Ruj, S.: A survey on security and privacy issues of blockchain technology. IEEE Commun. Surv. & Tutor. **22**(2), 196–209 (2018). https://doi.org/10.1109/COMST.2019.2933199
39. Kumar, A., et al.: R-PBFT: a secure and intelligent consensus algorithm for internet of vehicles. Veh. Commun. **38**, 100609 (2023). https://doi.org/10.1016/j.vehcom.2023.100609
40. Khan, M.A., et al.: A blockchain-based security architecture for vehicular ad hoc networks. Veh. Commun. **29**, 100334 (2022). https://doi.org/10.1016/j.vehcom.2020.100334
41. Yang, Z., Yang, K., Lei, L., Zheng, K., Leung, V.C.: Blockchain for vehicular networks: a survey on architecture, applications, security, privacy, and challenges. IEEE Commun. Surv. & Tutor. **21**(4), 3182–3213 (2019). https://doi.org/10.1109/COMST.2019.2916180
42. Ajakwe, S.O., Kim, D.-S., Lee, J.-M.: Drone transportation system: systematic review of security dynamics for smart mobility. IEEE Internet Things J. **10**(16), 14462–14482 (2023)
43. Jiang, T., Fang, W., Wu, Q., Li, J.: Blockchain-based internet of vehicles: distributed network architecture and performance analysis. IEEE Internet Things J. **6**(6), 4640–4649 (2019). https://doi.org/10.1109/JIOT.2019.2908699
44. Gebrezgiher, Y.T., Jeremiah, S.R., Deng, X., Park, J.H.: Machine learning-based blockchain technology for secure v2x communication: open challenges and solutions. Sensors **25**(15), 4793 (2025)
45. Eşer, M., Yılmaz, A.E.: Gossip based consensus algorithm for multi-agent systems. In: 2024 8th International Symposium on Innovative Approaches in Smart Technologies (ISAS), pp. 1–7. IEEE (2024)

46. Alrubei, S.M., Ball, E.A., Rigelsford, J.M., Willis, C.A.: Latency and performance analyses of real-world wireless IoT-blockchain application. IEEE Sens. J. **20**(13), 7372–7380 (2020). https://doi.org/10.1109/JSEN.2020.2979031
47. Chen, X.: Scaling byzantine fault-tolerant consensus with optimized shading scheme. IEEE Trans. Ind. Inform. **20**(3), 3401–3412 (2023)
48. Zhou, L., Diro, A., Saini, A., Kaisar, S., Hiep, P.C.: Leveraging zero knowledge proofs for blockchain-based identity sharing: a survey of advancements, challenges and opportunities. J. Inf. Secur. Appl. **80**, 103678 (2024)
49. Gong, C., Xiong, L., He, X., Niu, X.: Blockchain-based conditional privacy-preserving authentication scheme for vehicular ad hoc networks. J. Ambient. Intell. Humaniz. Comput. **14**(5), 6273–6286 (2023)
50. Bendiab, G., Hameurlaine, A., Germanos, G., Kolokotronis, N., Shiaeles, S.: Autonomous vehicles security: challenges and solutions using blockchain and artificial intelligence. IEEE Trans. Intell. Transp. Syst. **24**(4), 3614–3637 (2023)
51. Cui, M., Han, D., Liu, H., Li, K.-C., Tang, M., Chang, C.-C., Ayaz, F., Sheng, Z., Guan, Y.L.: Secure data sharing for consortium blockchain-enabled vehicular social networks. IEEE Trans. Veh. Technol. **73**(12), 19682–19697 (2024). https://doi.org/10.1109/TVT.2024.3448207
52. Huang, J., Kong, L., Wang, J., Chen, G., Gao, J., Huang, G., Khan, M.K.: Secure data sharing over vehicular networks based on multi-sharding blockchain. ACM Trans. Sens. Netw. **20**(2), 31–13123 (2024). https://doi.org/10.1145/3579035
53. Jain, R., Singh, A., Sharma, A., Park, J., Singh, A.: Blockchain and autonomous vehicles: recent advances and future directions. IEEE Access **9**, 131306–131338 (2021). https://doi.org/10.1109/ACCESS.2021.3113851
54. Leontaris, L., Mitsiaki, A., Charalampous, P., Dimitriou, N., Leivaditou, E., Karamanidis, A., Margetis, G., Apostolakis, K.C., Pantoja, S., Stephanidis, C., Tzovaras, D., Papageorgiou, E.: Low-latency PBFT parallel consensus for IoT blockchain platform. Futur. Gener. Comput. Syst. **116**, 316–329 (2021). https://doi.org/10.1016/j.future.2020.10.042
55. Rathee, G., Sharma, A., Iqbal, R., Aloqaily, M., Jararweh, Y.: On a blockchain-based security scheme for defense against malicious nodes in vehicular ad-hoc networks. Trans. Emerg. Telecommun. Technol. **30**(11), 3624 (2019). https://doi.org/10.1002/ett.3624
56. Yang, Y., Wei, L., Wu, J., Long, C., Li, B.: A blockchain-based multi-domain authentication scheme for conditional privacy preserving in vehicular ad-hoc network. IEEE Internet Things J. (2021). Cross-domain, pseudonym distribution, blockchain-based authentication
57. Ajakwe, S.O., Olabisi, K.L., Kim, D.-S.: Multihop intruder node detection scheme (MINDS) for secured drones' FANET communication. IET Intell. Transp. Syst. **19**(1), 70080 (2025)
58. Ajakwe, S.O., Kim, D.-S.: EQAI: Explainable quantum-empowered antispoofing intelligence for trustworthy connected autonomous vehicles communication. IEEE Internet Things J. 1–1 (2025). https://doi.org/10.1109/JIOT.2025.3638378

Architectures and Intelligent Trust Systems

Blockchain in AIV with Multiple Agents AI Architecture

Vadym Slyusar

Abstract The chapter proposes a systemic vision of integrating blockchain with the multi-agent architecture of autonomous intelligent vehicles in urban and non-urban settings. It shows how lightweight registries of events and policies sustain trust among sensor, planning, and executive agents without reliance on a centralized controller and without placing the blockchain on the hot path of motion control. It describes a two-circuit organization of accounting with an internal onboard AIV log and external permissioned domains that accept a compact state digest and regulate access on the basis of verifiable credentials and roles. It details coordination mechanisms via smart contracts, selective disclosure through commitments and zero-knowledge proofs, as well as support for device digital identity with DIDs and a hardware root of trust. It presents migration procedures between domains for ground AIVs, UAV swarms, and USV flotillas with history reconciliation and offline modes. Particular attention is given to loitering agents, which provide long-lived semantic memory of the environment, reduce latency, and balance load while operating at the Edge-Fog-Cloud boundary. The proposed architecture reconciles requirements for security, privacy, and scalability, enables transparent audit and interdomain interaction, and creates the technical foundation for reliable coordination of states and decisions in transport systems with a large number of heterogeneous agents.

Keywords Artificial intelligence · Blockchain · Multi-agent systems · Mixture of Experts (MoE) · 6G

1 Introduction

A key trend in the development of artificial intelligence (AI) technologies is the application of the mixture of experts (MoE) architecture [1–4] and its variations, as well as the transition to multi-agent systems (MAS) [5–7]. In this context, the design of

V. Slyusar (✉)
Central Research Institute of Armaments and Military Equipment of Armed Forces of Ukraine, Kyiv, Ukraine
e-mail: swadim@ukr.net

S. O. Ajakwe and D.-S. Kim (eds.), *Blockchain for Security of Autonomous Intelligent Vehicles*, Studies in Big Data 194,
https://doi.org/10.1007/978-3-032-20851-4_4

advanced autonomous intelligent vehicles (AIV) [8] should be oriented towards their operation in dynamic and multidimensional environments, where dozens or even hundreds of AI agents of different levels function simultaneously—from sensory and executive agents to cognitive agents of strategic planning. Each of them generates data, makes local decisions, and exchanges information with other nodes. Under such conditions, traditional centralized management models lose effectiveness due to the increasing risks of delays, unauthorized data modifications, and loss of synchronization.

To overcome these challenges, blockchain technology may be employed. As is well known, it enables the distribution of control and trust functions among all participants in autonomous collaboration, forming an immutable chain of transactions that reflects not only financial operations but also informational events, sensory measurements, and decisions taken. The concept of a hash chain for immutable event recording was first described in practice by Stuart Haber and Scott Stornetta in 1991 [9] as a method of time-stamping digital documents, where each new record is linked to the previous one through a cryptographic hash, thereby making substitution attempts visible. This construction relies on Merkle trees, introduced by Ralph Merkle in his cryptographic work of the 1980s [10], which provide a compact summary of large sets of records. The modern "blockchain" as a decentralized ledger with transaction blocks and proof-of-work was first systematically implemented by Satoshi Nakamoto in Bitcoin in 2008 [11], after which the term and the approach acquired their current meaning.

Given the prospects of blockchain technology, one of the key tasks is the organization of a multi-agent architecture for AIV [6], in which each AI agent acts simultaneously as both a consumer and a provider of data within the blockchain network. In a system of systems comprising multi-agent AIV of different purposes and the infrastructure with which they interact, blockchain emerges not as an auxiliary tool but as a fundamental infrastructural component ensuring trust, consistency, and security in interactions among agents.

The following section will examine in more detail the key features of integrating the multi-agent approach with blockchain-oriented principles in the construction of collaborative autonomous vehicles.

2 Features of Integrating the Multi-agent Architecture of AIV with Blockchain Technology

Regardless of the purpose of an AIV and the missions it performs, a multi-agent architecture requires a minimally necessary set of agents that provide the vehicle's functional capabilities. These include sensory, executive, and planning AI agents [6]. All of them may be implemented using neural networks such as CNNs, autoencoders, transformers, and both small and large open-source language models. Within an AIV, these agents form the vehicle's internal network similar to NGVA [12], distinguished by its reliance on specialized inter-agent data exchange protocols [13–17]. Externally, this network is connected through secure gateways to edge, fog, and cloud V2X infrastructures. At these gateways, the execution environment is verified, telemetry formats are aligned, and events are converted into standardized messages to ensure coordination with other traffic participants.

In the context of integration with blockchain technology, one possible implementation involves two interrelated yet distinct subsystems. Inside the AIV, an internal blockchain subsystem operates. It records events, the provenance of embeddings, policy changes, and local rewards. This subsystem functions autonomously and does not expose raw data externally. At the same time, intelligent sensory agents capture information about the surrounding environment and log it into the blockchain in the form of transactions, thereby ensuring transparent documentation of events. Planning agents provide navigation, generate routes and tactical movement scenarios, while executive agents verify their compatibility with local conditions. A representation of the described architecture is given in Fig. 1.

To the minimal set of agents, additional services may be added, including a user-interface agent capable of processing voice commands, maintaining dialogue, and responding to gestures. It is also advantageous to include a dedicated diagnostic and prognostic agent for monitoring and forecasting the state of the vehicle. For the purpose of cryptographic protection of the multi-agent system, it is reasonable to introduce a security agent responsible for controlling physical access to the vehicle on the basis of biometrics and for safeguarding language-model-based agents from undesirable queries and intrusion attempts.

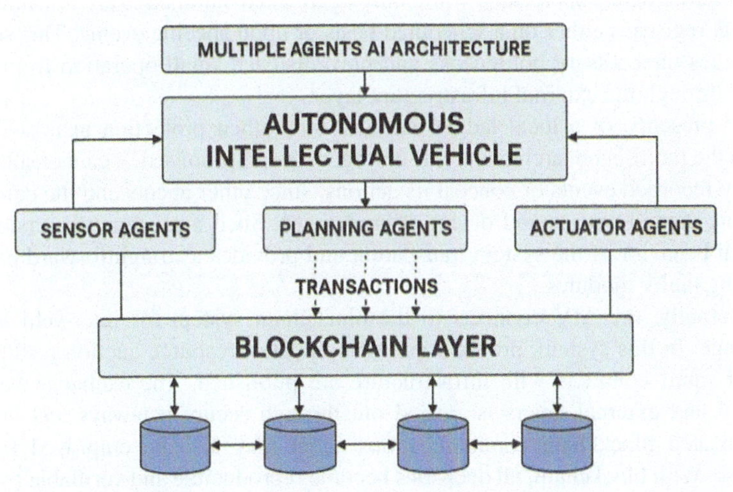

Fig. 1 Architecture of AIV with a multi-agent system and a blockchain layer (the image was generated using the OpenAI gpt-image-1 model)

The rationale for a dedicated blockchain ledger within an individual AIV lies in the fact that it provides the vehicle itself with an independent source of objectivity for all its agents. When navigation, sensory, diagnostic, and security agents operate in parallel, the onboard ledger orders their actions into a single chronological stream and prevents any agent from retrospectively altering an event. As a result, the multi-agent system attains internal consistency without continuous reliance on external networks.

When access to a city or domain network is temporarily unavailable, the AIV continues to operate, recording events with signatures and subsequently anchoring the summarized record in the municipal or transportation registry. This mode ensures that the reproducibility of history is preserved and that critical control functions are not suspended.

The AIV's internal ledger also functions as a layer of trust and accountability. The vehicle stores minimal metadata with hashes and time stamps, binds them to its DID identifier, and signs them using keys from a hardware security module. A DID is a decentralized identifier that allows an entity to possess a persistent name and its own cryptographic keys without a central registry. The DID owner controls the keys and may operate with them without altering the identifier itself, while interacting parties can verify signatures. This provides verifiable identity for devices and agents without reliance on a specific provider.

The recorded trace in the registry thus enables proof that a particular vehicle has executed specific actions, while external domains receive compact proofs without disclosure of unnecessary data.

Another advantage of a local ledger is the reduction of latency and traffic. The AIV makes minor decisions locally, while only aggregated summaries are transmitted to external registries either on a scheduled basis or upon specific events. This scheme eliminates unnecessary bottlenecks and prevents each small operation from being forced through the external infrastructure layer.

The presence of a local ledger can also strengthen protection against misuse within the multi-agent architecture. If one agent is compromised, it cannot substitute already recorded events or conceal its actions, since other agents and the controller compare signed entries and detect discrepancies. Such a mechanism renders the internal behavior of the system transparent and provides a straightforward path for isolating faulty modules.

Externally, the AIV connects to the blockchain system for inter-vehicle data exchange. In this system, proofs, reputation updates, resource auction results, and agreed smart contracts with infrastructure are published. The exchange between internal and external layers is carried out through secure gateways and bridges. A dedicated blockchain communications agent may also be employed for this purpose. With blockchain, all decisions become reproducible and verifiable by other participants, thereby eliminating the issue of conflicting or duplicated decisions.

To avoid transmitting sensitive data, commitments and zero-knowledge proofs are exchanged instead. This preserves privacy, ensures trust, and enables inter-vehicle collaboration.

An important aspect is the use of smart contracts as mechanisms of automated coordination. They can govern, for example, traffic priorities at intersections, rules of interaction with infrastructure, or algorithms for data exchange between vehicles based on 5G and 6G technologies [18, 19]. Such logic operates in real time and does not require an external controller, which is critically important for high-speed scenarios on roads.

Another benefit of blockchain in AIV is the provision of cybersecurity not only for internal but also for external networks. Distribution and cryptographic resilience make the system resistant to manipulations by individual nodes. This is reinforced by the multi-agent nature of the system: different agents perform independent verification and confirmation functions for transactions, thereby creating a multi-layered trust-based interaction architecture.

A particular significance is acquired by the integration of blockchain with methods of decentralized learning. In such a configuration, agents may not only share complete models but also enter into the blockchain metadata concerning parameter updates, quality verification, or test results. This creates the possibility of collective evolutionary learning of the transport network without the need for a centralized server, while preserving the transparency and reliability of the process.

Blockchain also provides a foundation for resource management. In conditions of multichannel communication, when an AIV simultaneously employs several types of sensors and communication channels, smart contracts may regulate the prioritization of access to network resources, control energy consumption, or balance loads among computational nodes. This forms a resilient system capable of self-organization and scalability.

In the longer term, the application of blockchain in multi-agent architectures of autonomous vehicles can ensure a new level of interoperability. Transport systems from different manufacturers, cities, or even countries may obtain a unified interaction protocol, where all transactions are verified and trust is guaranteed by consensus algorithms. This opens the way for the formation of a global network of autonomous transport functioning in a mode of mutual coordination without a centralized regulator.

To enable an AIV to react instantly to environmental situations, the blockchain must not lie on the critical control path of the vehicle. The system separates three layers: real-time for motion control, a local ledger for evidential record-keeping, and background notarization in domain registries. The real-time layer operates entirely locally under the supervision of a security processor and does not await any confirmations from the network. Agent decisions and sensory events pass through the local event bus, where each message immediately receives a hardware-timer timestamp and a computed hash. The system then exports all actions into the slower layer: it packs events into micro-batches and periodically anchors only roots and compact proofs into the domain registry, without blocking control functions. If the network is unstable, the background queue applies micro-batching with an upper bound, for example, up to 50 ms or N events, and never interferes with real-time priorities.

To avoid possible delays in authorization and payments, the node must be prepared in advance. The vehicle and agents operate under pre-issued short-lived credentials and resource budgets with defined limits and expiration times, so verification on the critical path is reduced to local validation of signature and validity. Payments and tariffs do not pass through the chain at the moment of action but are deducted from pre-opened channels or deposits and recorded afterwards as part of aggregated summaries. If the limit is exhausted, the system lowers the priority of paid services but does not block essential safety functions.

To ensure that event logging does not introduce jitter into the AIV control system, computational isolation is mandatory. The security layer and minimal inference operate on dedicated cores with fixed time budgets. Agents, including those based on LLMs, are executed in separate containers or Trusted Execution Environments (TEE) with ceilings on FLOPS, memory, and input–output processes. In high-load scenarios, degradation affects inference quality but does not touch the control layer. Hashing and signatures are accelerated by hardware. As a result, the vehicle responds as quickly as a system without blockchain, because blockchain is not involved in the decision-making loop. This allows millisecond-level motion safety to be combined with transparent auditing, inter-domain roaming, and long-term verifiability.

Thus, the integration of blockchain with multi-agent AIV architectures creates a multidimensional infrastructure in which decentralization and collective intelligence are interconnected. This represents not only a technological innovation but also a foundation for constructing new socio-technical systems, in which vehicles become full participants in the digital economy and networked society.

In the following, we shall analyze specific aspects of this approach in greater detail.

3 Hierarchical Blockchain Architecture

In the future quantum–classical infrastructure of a smart city, different multi-agent networks will coexist. This raises questions about the role of blockchain and whether it is better to adopt hierarchical blockchain architecture or a single blockchain network for the entire city.

At first glance, a single unified blockchain network may appear simpler, but it quickly becomes a bottleneck. Such a choice rigidly couples the updates of all multi-agent subsystems, complicates the enforcement of distinct privacy regimes for energy, transportation, healthcare, and security, and worsens latency for edge-level scenarios [20]. In contrast, a hierarchical blockchain architecture with several domain-specific registries ensures local performance and fault isolation while allowing the implementation of diverse access policies and cryptographic schemes.

The registries within multi-agent domains are physically replicated across multiple validator nodes. This follows the classical model of state replication: all validators store a full or partial copy of the hash chain and reconcile each event via a consensus protocol. As a result, all nodes obtain the same outcome, even if some are compromised or lag behind. Thus, "a single registry" refers to one common version of the true system of records rather than a single server or instance of data.

Meanwhile, a thin urban trust layer accepts only a compressed summary of the state of subordinate multi-agent networks. In practice, this means that for transportation, energy grids, healthcare, public safety, and the urban Internet of Things (IoT), it is appropriate to maintain separate permissioned registries with BFT-class consensus (Byzantine Fault Tolerance), which confirm operational events with millisecond- or second-level finality at the edge. BFT-class consensus represents a family of state-agreement protocols for distributed systems that preserve correctness even when part of the nodes behave arbitrarily or maliciously [21]. Such nodes are referred to as Byzantine, and the classical resilience threshold for most practical algorithms is expressed as $n \geq 3f + 1$, where n is the number of participants and f is the maximum tolerated number of faulty or compromised nodes. In simple terms, a system of 4 nodes can withstand 1 traitor, a system of 7 can withstand 2, and so forth, while uncompromised nodes still converge on the same final decision and event order.

In addition, into the urban blockchain chain, every N seconds or minutes aggregated commitments are published in the form of Merkle roots or zero-knowledge proofs. A Merkle root is a compact cryptographic label summarizing a large set of records in such a way that any individual record's membership can be efficiently verified. The system takes each record, computes its hash, then pairs neighboring hashes and re-hashes them, repeating the process up the tree until one final hash remains at the top. This is the root. For example, if records e1, e2, e3, e4 have hashes h1, h2, h3, h4, then the intermediate nodes of the next hierarchy level are $h12 = H(h1 \| h2)$ and $h34 = H(h3 \| h4)$, and the root is $r = H(h12 \| h34)$. To prove that e3 is included in the summary, it is sufficient to provide a short proof path consisting of several neighboring hashes. If even a single byte in any record is changed, the root will change, and the tampering will be evident. The corresponding verification takes time proportional to the logarithm of the number of records, rather than the size of the entire set. In scenarios of interaction between multi-agent AIV architectures and domain registries, the Merkle root is a convenient way to publish a compressed session summary or a group interaction history in the appropriate domain.

The advantage of this approach is that inter-domain transactions do not block local processes, fault isolation prevents cascading failures, and scaling to city districts is achieved by adding new chains without redesigning the entire solution. The upper layer of the hierarchical blockchain architecture functions as a dispute arbiter, characterized by low bandwidth requirements but high requirements for cryptographic resilience and governance.

Naturally, the hierarchical approach has an evident drawback: more complex inter-network interaction, which requires formally verified bridges, standardized data schemas, and a carefully defined trust model for event relays.

Given the anticipated coexistence of classical and quantum networks at all levels in the future, it is recommended to apply hybrid signatures combining post-quantum algorithms with classical elliptic curves during the transition. Moreover, it is necessary to maintain updatable cryptography profiles at the domain level and canonical policy catalogs specifying permitted algorithm sets and key sizes. Quantum communication is best suited for distributing session keys between datacenters or critical nodes, while the signing mechanism for transactions should rely on post-quantum cryptographic schemes, since signatures themselves are the primary targets for storage and later compromise. For digital identity, verifiable credentials with self-sovereign identifiers can be conveniently applied, where different city domains issue attributes limited by scope of authority, while the urban blockchain acts as a source of public trust and revocation registries.

For inter-domain operations, it is advisable to standardize the event model and asset types so that atomic exchanges can be executed via bridges with state proofs, without the need for strict synchronization of all ledgers. For emergency scenarios, rollback plans are required, enabling local domains to function autonomously and resynchronize with the city anchor after connectivity is restored, with all bridges protected against replay and delayed messages.

If the city is small and the case concerns a pilot project, one may begin with a single permissioned chain with clear namespace separation and contractual domains, but from the outset, the scheme should be designed for splitting into multiple domain chains to ensure a seamless migration.

For megacities or environments with stringent privacy requirements and low latency, it is better to adopt a hierarchical model immediately. Such an approach enables functionality expansion without interrupting the city's operations and establishes clear boundaries of responsibility, which is particularly important for real-world risk management and regulatory compliance.

4 Robotic Road Transport

In the case of an urban robotaxi, the use of blockchain for payments is only one among many promising scenarios and is usually not the primary one. First and foremost, blockchain serves as the basis for the digital identity of the vehicle and its AI agents, since the system stores verifiable credentials. This makes it possible to issue short-term passes for lanes, parking areas, and service zones without disclosing unnecessary information about the user. In addition, municipal services use blockchain to notarize policies and access rules, which simplifies compliance verification and ensures an unbroken chain of custody for sensor data and video. Given the recording in the blockchain of immutable time stamps for the actions of the vehicle and its agent subsystems, responsible parties can reconstruct the sequence of events after an incident without disputes. At the same time, the system enables transparent accounting of resource usage, for example, charging stations, toll roads, or access to APIs, and the accounting is not necessarily tied to money. In cooperative scenarios, including during platooning of autonomous logistics convoys with coordinated manoeuvres, participants can unambiguously verify the fulfilment of agreements because blockchain records commitments. The same applies to other swarming scenarios.

Under federated training of agent models, when several robotic vehicles learn joint manoeuvres, only hashes and checksums of update packages are retained, which helps with advisories and recalls through public revocation lists of certificates and components. Furthermore, blockchain confirms the provenance of software and firmware updates for their multi-agent versions, whereby the system protects the vehicle against substitution and unauthorized modifications as an anti-spoofing measure.

5 Procedures for Migration Between Domains and Localities

If one considers a robotic automobile with its own blockchain in its multi-agent architecture, then in this case, a multilayer hierarchy of blockchains is possible. Let us examine how, within it, the migration of the automobile between adjacent domains of the transport blockchain architecture, or even from one city to another, should be handled. For a robot mobile with its own internal ledger to transition safely between the transport domains of a city, it is convenient to conceive the corresponding migration as a short-term change of trust context with atomic fixation of state. At the same time, it should be taken into account that in a smart city there will be several major blockchain registries for different multi-agent spheres, for example transport, energy, and security, above which a lightweight municipal notary operates that only attests key facts. When a vehicle leaves one transport domain and enters another, this resembles crossing a border with electronic documents. Before departure, the vehicle makes a control mark in its onboard journal, closes open accounts and obligations, summarises them into a single short entry, and signs it with its hardware key. The old domain records that all matters are closed and issues an electronic exit ticket. At entry into the new domain, the vehicle presents this ticket and a short proof of its previous state, without disclosing the driver's personal data or the full route. A dedicated agent of the new domain verifies the ticket according to simple rules, grants temporary permissions for movement, parking, charging, and other services, and informs the municipal notary that the automobile has changed domain. In this way, the municipal layer sees important events but does not store detailed daily logs.

If the automobile relocates to another locality in which a blockchain system is also supported, the principle remains the same, except that the role of the notary is performed by an AI agent at the intercity or regional levels. Payments are conveniently transferred as rechargeable electronic wallets or travel passes, which are locked in one domain and unlocked in another without conversions. Access to priority lanes or service zones can be issued as short-term electronic passes, where the automobile proves that it meets the requirements but does not disclose unnecessary information. For privacy, the AIV periodically changes public identifiers, while the true binding to a specific vehicle is confirmed by a protected hardware module and technical records from service stations.

If, in the process of changing location, connectivity disappears, the automobile continues to operate autonomously, records all events in its onboard journal with numbering and a hash chain, and attempts to perform a minimal exchange with a roadside node or a border gateway that can accept a short offline attestation of state. When this is not available, the domain permits exit by an offline pass with restrictions, for example, to the nearest connectivity area or service zone, and specifies a time window within which synchronization must be completed. At entry into the new domain, the automobile can present the previously signed state and obtain temporary permissions with small limits. As soon as connectivity appears, reconciliation of history occurs according to the rule of the largest finalized prefix, all offline events are verified and extended into the chains of the domains, and payments are processed with updated rates and tariffs if this is provided by policies. Duplicate messages and double spending are culled by signatures and event numbers, and in the event of discrepancies there is a simple dispute mechanism with clear timers.

From the standpoint of protection, at the design stage of the described system it is better to embed hybrid cryptography at once, with algorithms that will remain robust even after the emergence of quantum computers. The data of people and detailed telemetry records should not be stored in the chains. They should contain only hashes, time stamps, and access policies, while the data themselves are advisable to keep in managed repositories with controlled access.

The scheme under consideration allows the automobile to migrate freely between districts and cities without excessive bureaucracy, with low latencies for local operations, clear boundaries of responsibility, and an intelligible picture for risk control. If, upon exiting a domain, the robo-automobile cannot close accounts either due to insufficient funds or due to lack of connectivity, there are two basic scenarios, and both must operate safely and predictably. When funds are lacking, the domain does not block the automobile in the middle of the road but switches it to a soft-exit mode. By default, a small deposit is held at entry and an offline spending limit is set. At exit, the system first attempts to top up the wallet automatically from a linked source, if this fails it debits from the deposit, if this is also insufficient it issues a short-term pass to the nearest top-up point or service zone. All outstanding charges are folded into a single signed debt record with a redemption term visible to audit. After successful top-up, this record is extinguished and a normal exit checkpoint is formed. If the debt is not closed within the allotted time, a pre-agreed rule is triggered, for example partial deduction of collateral, lowering of priorities for access to paid services, or a prohibition on new paid operations until settlement. Critical safety functions such as movement on main roads and the ability to park are not cut off, but paid services, in particular passage on high-speed lanes, use of paid parking, or charging, may be temporarily unavailable until settlement of accounts.

All the specified procedures are aimed at guaranteeing traffic safety and the integrity of accounting, as well as bringing financial and access matters to completion as soon as there is an opportunity to replenish funds or perform synchronization.

6 Aerial Drones

When, in addition to robotic automobiles, aerial drones are used within a smart city or beyond it, a dedicated aviation registry domain must be created for them inside the city hierarchy. The corresponding aviation blockchain should exist alongside the considered transport domain for ground vehicles with a shared lightweight trust layer at the citywide level. This is explained by the fact that airspace is subject to different safety and oversight rules than road traffic. Drones need to have flight plans, Remote ID markings, take-off and landing permissions, geofencing commands, and logs of loss of communication and incidents. For ground vehicles these data are superfluous, and their appearance in a single shared registry creates noise and excessive load. A separate aviation domain allows the application of its own access policies and risk model and its own mode of event finalization without blocking other systems.

The structure of the blockchain system of a multi-agent aviation domain should be as follows. Each drone maintains an on board immutable log of events and states as a lightweight local chain. The city's aviation domain operates as a permissioned registry with BFT-class consensus and accepts from drones compact anchor records with hashes and time stamps. The city's lightweight trust layer accepts only aggregated summaries and policies so that different domains can interact with each other. Bridges between domains allow a drone to obtain temporary passes for landing at transport hubs or for shared use of charging infrastructure without disclosing unnecessary data.

The system must separate flight safety channels from the blockchain, therefore collision avoidance commands and real-time telemetry must go over direct channels with minimal latency, while only hashes and time stamps for audit are recorded in the registry. The identity of a drone must be built on a hardware root of trust and verifiable credentials, and public pseudonyms are periodically rotated for privacy. Large data, for example video or raw sensor streams, similarly to the automotive segment are stored off-chain in managed repositories with attribute-based access, while only hashes, policies, and integrity proofs should be in the chain.

Roaming between districts and cities must function in the same way as the described option for automobiles. A dedicated agent of the drone forms a checkpoint of exit in its on board log, the agent of the aviation domain closes open obligations and records a short proof of state, and the new domain accepts this proof and issues short-term permissions. If there is no connectivity, the drone switches to an offline mode with limited permissions and synchronizes at the first opportunity. Financial settlements for landings, corridors, or aeronautical services may be conducted through a payments subdomain, but the aviation registry itself is not reduced to payments; it serves primarily as an anchor of trust, audit, and policy consistency.

Such a design gives drones their own rules and operating tempos, preserves the shared city trust layer for inter-domain operations, and does not overload transport systems with unnecessary events. It is simpler to maintain, scales better, and transparently passes flight safety audits.

7 Multi-swarm Architecture of a System of Systems of Drones

In the case of several drone swarms that interact and intersect in the air (Fig. 2), it is advisable to make the blockchain multi-domain and lightweight so that it does not stand in the way of flight control. In such a configuration, each swarm maintains its own onboard operational log as a permissioned chain with short entries about sessions, policies, Merkle roots from event summaries, and key versions. Above them there exists a municipal or regional aviation domain in the form of a trust layer to which the swarms periodically submit a compressed summary. Communication bridges between swarms operate as light clients without trust in relays, which merely verify state proofs and are not permitted to alter history. Information for real-time control, collision avoidance, and telemetry is routed over direct low-latency channels, while only hashes, time stamps, session identifiers, and brief proofs of policy compliance enter the chains, as in the cases considered.

A cross-flight involving the passage of one swarm through another is organised as a short joint session with two classes of artefacts. The first artefact is a corridor manifest recording geometry and time windows, the maximum number of vehicles per layer, priority rules, and requirements for sensor state. The second artefact is access tokens issued by the aviation domain or an authorised dispatcher that confirm the right of specific drones to enter the corridor in defined slots. Both swarms sign the manifest, each publishes to its chain the Merkle root of its movement plan, and in response receives short-lived tokens bound to their DID and session number. To avoid disclosing the number of vehicles and the routes, a swarm provides a proof that the aggregate number of drones in each time layer does not exceed the limit

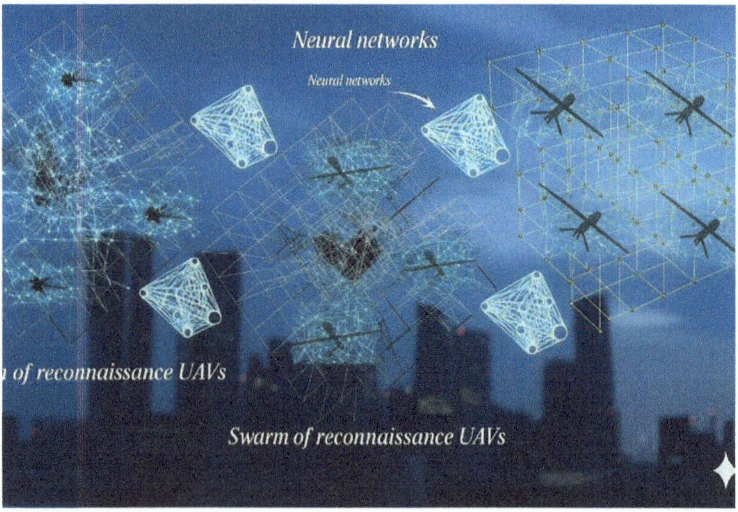

Fig. 2 Swarm of drones (the image was generated using the Google Nano Banana model)

and that all drones have a valid technical condition and Remote ID. Only the fact of compliance appears in the chain, without transmission of routes or video streams.

During mutual penetration the swarms do not exchange streams of raw data unless necessary. They send one another only brief receipts for corridor entries and exits with precise time stamps from onboard timers and with signatures of hardware security modules. Each such receipt includes the hash of the onboard event, the session number, and a reference to the joint manifest. If a station or satellite beacon witnesses the intersection of the swarms, it signs a third receipt as an independent observer. These triangular confirmations render the protocol self-sufficient for audit yet lightweight for the network. Aggregated roots are published to the aviation domain no more often than once every few seconds, which does not affect control.

Although each swarm and each drone has its own DID, in cross sessions it makes sense to assign short-lived pseudonyms to drones, while the linkage to the actual device is proven by a hardware certificate in the event of an incident. Conflict and priority resolution operates without interrupting flight. If a swarm does not manage to vacate a layer in time, it automatically reduces density, confirms this with a brief event, and receives a new slot from the dispatcher without revalidating all policies. If part of the swarm switches to silence mode, the drones operate under offline passes with limited rights, number events, and publish deferred summaries as soon as a channel appears. In the case of a dispute, it is sufficient to present for arbitration the signed entry and exit receipts, the Merkle path to them, and a reference to the manifest.

Joint tasks between swarms such as relaying, search [22], or logistics are formalized as delegation of authority. Swarm A issues to Swarm B a short-term capability for relaying or surveying a variable area, with boundaries and budgets specified. Swarm B signs acceptance and shares heartbeat signals, and each act of exercising the rights is recorded as a micro-event with a hash. Upon completion of the joint mission the delegation is automatically revoked, and the final root appears in both logs and in the aviation domain. Such symmetry does not require a central dispatcher but makes double use of the same rights impossible.

8 An Aerial Drone as an Individual Auxiliary Means of a Motor Transport Platform

When a drone is mounted on a vehicle and used in its interests as an individual instrument, such a drone, quite obviously, must be an integral component of the vehicle's blockchain. At take-off the drone will migrate into the blockchain of the air domain in such a way as not to lose connection with its vehicle and its registry. The situation should be formalized as dual subordination of the drone with clear session boundaries. In this case entry into the aviation domain is advisable to formalize in the status of a temporary participant with special authorities. This is achieved not by duplicating all accounting, but by neatly linking the layers of events through short state proofs and delegated rights.

For identity determination the vehicle has a root DID with a hardware root of trust, and the drone inherits from it a child DID with its own keys. The vehicle issues the drone verifiable credentials and a short-lived delegated capability for flight, where the area, altitude, time, and available services are strictly constrained. This creates a cryptographic linkage "the vehicle as sponsor, the drone as executor" without disclosing excessive data to external parties.

During the pre-flight handshake the vehicle and the drone create and sign a flight session identifier, as well as the first anchor record for the vehicle's onboard chain and for the aviation domain. On the transport side there appears a short commitment "the drone is detached, session X is active," and on the aviation side there appears a record "drone X has joined with delegated authorities." Flight control, collision avoidance, and telemetry go outside the blockchains over low-latency channels.

After take-off the primary log of flight events is written into the aviation domain in compliance with its rules and finalization. At the same time the drone periodically sends into the vehicle's onboard chain a compressed summary of the growing log in the format of a Merkle root with event numbers. In this way the vehicle does not lose the drone's linkage to its own accounting and can prove that it was precisely this drone that acted on its behalf within the assigned authorities. Cross-anchors with the same session identifier ensure reproducibility of the history without duplication of large data.

When the connection with the vehicle disappears, the drone does not stop the flight. It continues to publish events into the aviation domain, and for the vehicle's onboard registry it keeps a local offline log with numbering and signatures in order to build the missing anchors immediately after restoration of the channel. If there is no connection even with the aviation domain, the drone according to the mission tasks flies in offline mode with minimal permissions and is obliged to synchronize within the nearest corridor of time. Duplicate events and attempts at double accounting are filtered by numbers and signatures, and disputable situations pass through a simple dispute mechanism with timers.

After landing, the drone performs closure of the current session. In the aviation domain, there appears a finalized flight summary, and in the vehicle's onboard chain there appears a corresponding anchor record with the same session identifier. Small financial settlements for landing or corridors, if they occurred, are closed as a separate payment subevent, but the logs themselves remain domain-specific. As a result the entire history from take-off to landing is chain-linked and subject to audit both from the side of the urban air domain and from the side of the vehicle.

Security and privacy in this case are supported by several simple rules. The drone's pseudonyms are rotated between sessions, but the binding to a specific airframe is proven by hardware certificates and delegation digital signatures. Signatures are preferably made hybrid with support for post-quantum algorithms in order to withstand a long-term threat horizon. Bridges between registries verify state proofs as light clients and do not trust individual relays.

In the described way, the drone can function simultaneously in two domains. For the duration of the flight it acts as a full participant of the aviation segment of the blockchain architecture, and through periodic compact anchors it preserves continuity of the history in the vehicle's onboard registry. Such linking minimizes duplication, does not affect the latency of control, and provides a clear, reproducible picture for safety, audit, and incident resolution.

9 Aerial Drones and USV

In addition to the considered variants, surface unmanned platforms (USV) of a multi-agent architecture also merit attention, for which the principles of blockchain support described above remain in force. It should be noted that the multi-agent architecture of USV may have a hierarchical structure in which certain segments constitute multi-agent clusters. In particular, such a cluster may be formed by sensory agents, each of which services a separate sensor onboard the USV (Fig. 3). The list of radio-frequency and optical sensors presented in Fig. 3 may be supplemented by sonars, GNSS signal receivers, acoustic sensors, and others. Blockchain is also conveniently extended to the communication between USV and UAV, in particular when a relay drone services a group of unmanned boats in a network (Fig. 4). In this case, each USV maintains its own onboard event log in the form of a lightweight local hash chain, while in the maritime domain, a permissioned registry for flotillas operates. Upon take-off, the UAV creates a flight session and joins the aviation domain, but at the same time, it receives from the flotilla delegated authorities for relaying and notarizing their events. Owing to this, the drone maintains two logical contexts without duplication of large data and without loss of linkage with the vessels.

Before the start of the mission, the flotilla forms a group session with an identifier that includes all USV and the UAV as a trusted relay. The flagship USV or a ground operator issues short-lived credentials and an access policy that define which events the drone must record, how frequently it must publish aggregated summaries, and which zones it is permitted to use as air corridors. The UAV signs acceptance of these terms and publishes the first anchor record into the maritime multi-agent domain as proof of readiness to service this specific group. Thereafter telemetry and collision-avoidance commands are transmitted outside the blockchains over low-latency channels, while the registries receive informational records containing hashes, timing marks, and simplified attributes confirming the integrity of the chain.

During operation, the relay collects events from each USV, classifies them by type, and a separate agent within it constructs for the group a growing summary in the form of a Merkle root with numbering. The UAV periodically publishes this summary to the maritime domain, which provides an external verifiable trail for audit and incident resolution. In parallel, the UAV sends a compact digest to the onboard chains of each USV to preserve the continuity of their own histories without creating excess traffic. If some vessels, according to the mission plan or due to interference,

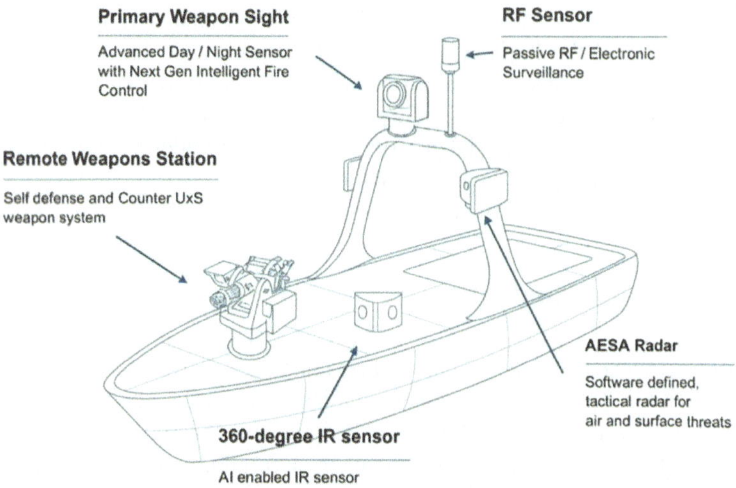

Fig. 3 The set of AI agents of the sensor cluster of the multi-agent hierarchy within the USV (the image was generated using the Google Nano Banana model)

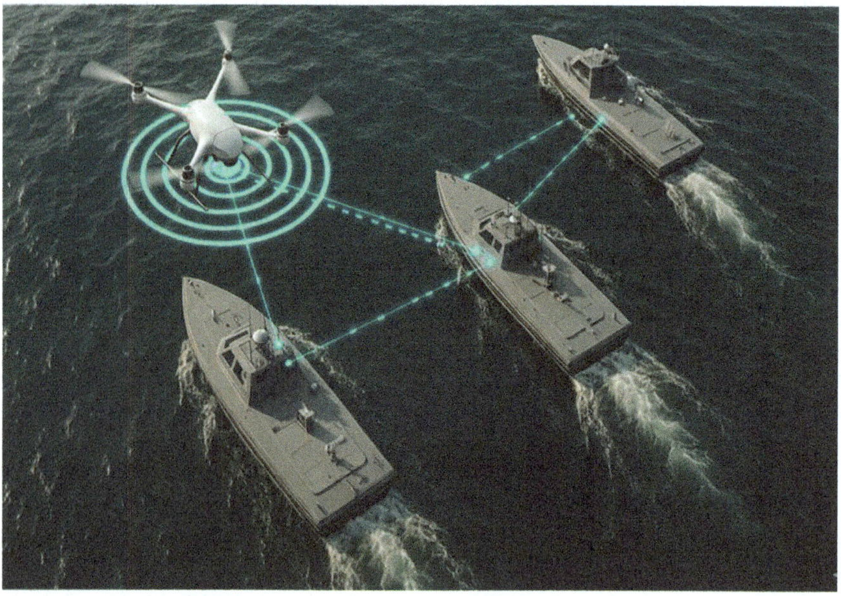

Fig. 4 Mesh Networking based on UAV + USVs (the image was generated using OpenAI gpt-image-1 model)

switch to radio silence, the UAV retains offline receipts for them and delivers these at the first opportunity, blocking possible duplicates of signatures and event numbers.

In the event of a sudden loss of connection with the maritime domain, the entire system smoothly transitions to offline mode. The UAV continues relaying within the group and maintains a local log with strict numbering and signatures. Each USV stores its events in its own chain and accepts temporary acknowledgements from the relay. After the channel is restored, the UAV reconciles the history according to the rule of the largest finalized prefix, publishes the accumulated aggregates, and obtains from the domain confirmation that no conflicts exist. If conflicts arise, the system applies a simple challenge procedure with timers, where any participant may provide their own state proof to resolve the dispute.

Security and privacy do not depend on the role of the payment layer. Payments for provided movement corridors, berths, or service operations may proceed in a separate subdomain, but the principal meaning of the logs in this scenario lies in trusted accounting of actions, access rights, and policies. The identities of all autonomous platforms are likewise built on a hardware root of trust, and pseudonyms are rotated between missions, so external observers do not obtain excessive information about the group's interaction patterns.

Mission completion occurs under the same conditions of predictability as the start. The UAV publishes the final aggregate of the group session in the maritime domain and attaches the same session identifier to the short digest in each USV's onboard chain. Thereafter the group authorities are automatically revoked, and new missions commence with new keys and policies. If the same drone services the next group, it first closes the previous session and then repeats the short joining procedure so as not to mix logs or create unnecessary trust linkages.

As a result, the same rules of dual affiliation and session thinking that apply to the linkage of a vehicle with drones operate unchanged for USV configurations with a UAV, including the mode of one or several relays for a group of vessels. This yields transparent reproducibility of events, low control latency, and a simple audit scheme without excess load on the network infrastructure.

In perspective, with the formation of digital ocean architecture, blockchain technology can be scaled analogously to the underwater environment as well, in particular for servicing seabed critical infrastructure. In the context of AIV, one should, in this case, consider combinations of UAVs and USVs together with underwater vehicles (UUVs) [23], which may operate autonomously or in collaboration with USVs (Fig. 5).

In this case, in addition to the considered blockchain networks, an underwater multi-agent domain must be established with acoustic communications for data transmission, for example, based on the JANUS communication standard (NATO STANAG 4748) [24–26].

Fig. 5 Collaboration of UAV + USV + UUV (the image was generated using the OpenAI gpt-image-1 model)

10 Loitering Agents as an Object of the Blockchain

The concept of integrating blockchain technologies with AIV considered above is based on a stable multi-agent architecture of vehicles. At the same time, as a promising direction for the development of MAS, the concept of loitering AI agents is proposed, which are an analogue of AIV in cyberspace. It presupposes that AI agents are not permanently assigned to a single node but are capable of dynamically moving between different computational environments, orienting themselves to specific conditions such as the nature of the load, the availability or reachability of required data, the level of confidentiality, and the permissible energy expenditures. They may change their location in the vertical hierarchy of Edge, Fog, and Cloud, as well as move horizontally between peer systems. The basis of such mobility is the ability of agents to transport their weight coefficients and internal state together with a containerized microservice environment.

In the context of the physical infrastructure of AIV, loitering models are best understood as a strategy of non-stationary spatio-temporal residence and computational presence, whereby agents migrate or are deliberately retained within certain zones or routes, maintain local embedding spaces, accumulate context, and perform tasks in the immediate proximity of data sources. Such behaviour of agents provides several basic advantages for a fleet of AIV. First, decision-making latency is reduced, since the transformation of sensor streams and the primary inference are performed at the network edge near the events. Second, resilience to connectivity interruptions increases, because a loitering agent can act autonomously, maintaining local models and caches of embeddings and, after the channel is restored, reconcile state with the cloud or neighboring nodes. In addition, the requirement for bandwidth decreases, because instead of raw data streams, the network transmits compressed semantic contents in the form of embeddings and aggregated features.

The practical usefulness of Loitering agents manifests itself, for example, in tasks of traffic coordination. Loitering agents that patrol intersections or critical segments of transport infrastructure in a smart city accumulate local reference embeddings for typical objects and trajectories of traffic participants, update risk maps, and disseminate the most useful representations to other AIV that pass through the zone only briefly. This reduces cold start for new participants, increases the probability of early anomaly detection, and shortens the time for re-orientation in a complex environment. In scenarios of platooning or evacuation, loitering nodes assume the role of temporary relays of embedding and decisions, helping to balance the load among AIV of different computational capabilities and resolving local conflicts of priority without recourse to a central dispatcher.

Resilience to nonstandard conditions increases because loitering agents form a long-lived semantic memory of the area. They are capable of tracking slow shifts in event distributions, for example, changes in typical trajectories of pedestrians, the emergence of new obstacles, small alterations in infrastructure, and of adjusting local planning rules. Such accumulation of context is difficult to reproduce under purely transit presence. In addition, loitering models naturally fit into decentralized reinforcement learning, where rewards for useful data or successful prevention of incidents can be accrued using Fog-level oracles [27], and the events themselves are recorded in a trusted ledger for subsequent audit.

An important component of usefulness is data and privacy management. The loitering mode allows selection and aggregation of information on Edge devices, deletion of redundant data, and retention of sensitive fragments within the perimeter while disseminating only proofs of compliance or hashed references.

From the standpoint of resource management, loitering models provide a better balance of energy and computation. An agent that remains in a zone can optimally plan cycles of activity and sleep, use caches of models, share load with neighboring nodes, and participate in micro-auctions of resources for peak events. This lowers the total cost of operation and increases the uniformity of system performance, since each transit AIV does not have to deploy heavy models from scratch or transmit large volumes of raw data.

In a case where agents of different manufacturers or organizations are on duty in a shared zone, they can exchange indices of semantic artifacts and maps of correspondences between different embedding spaces. This ensures coordination in multi-vendor and interdepartmental networks. At the same time, it is important to achieve interoperability on the basis of a unified standard of models, which is critical for scaling AIV to the level of a city or region.

The practical application of loitering agents covers urban intersections, industrial territories, logistics hubs, areas of natural disasters, and theaters of military operations. For example, in a smart city, loitering agents that are localized as needed onboard AIV provide a service of cooperative surveillance of intersections, reduce the probability of collisions, and help to quickly bypass obstacles. In manufacturing they maintain stable embedding indices for robotic loaders and inventory drones. In humanitarian operations, loitering zones become nodes for the accumulation of evidence about the state of infrastructure and evacuation routes, where it is important not only to make decisions here and now but also to ensure the reproducibility of the history of events.

To realise the indicated advantages, it is advisable to follow several engineering principles. First, plan the geography of loitering with regard to coverage by sensor channels, expected traffic, and risk zones, ensuring overlap of observations and robust bypass routes. Second, construct two-layer pipelines, in which fast local actions and search operate at the edge, while slower reconciliation, anchoring of indices, and audit are performed at Fog levels and in the Cloud.

To eliminate the impact of agent migration processes on the functioning of AIV, decisions regarding the relocation of agents are taken locally. An agent that is migrating before exit forms a final hash of its micro-log and obtains a one-time capability for launch at the destination point. After launch in the new location, the agent publishes the first anchor on the new node and continues operation. If publication of the anchor is temporarily impossible, the node stores the receipts locally and uploads them in the background, so migration does not create blocking.

The implementation of agent loitering in a certain way increases the load on data transmission channels. To mitigate this problem, the communications infrastructure should rely on 5G or 6G technologies [18, 19]. However, in addition, other approaches may be involved, consisting, for example, in spectral densification of communication channels on the basis of non-orthogonal methods of frequency multiplexing of harmonic signals [28–30], as well as the non-orthogonal densification of signals Orthogonal Time Frequency Space Modulation (OTFS) proposed in [6, 31, 32]. This will make it possible to increase the information capacity not only of radio links for agent migration via VANETs [33] and Vehicle-to-Infrastructure Networks [34], but also of laser and acoustic channels of data exchange between the AIV themselves and the surrounding infrastructure in different physical environments.

Overall, loitering models provide AIV with the ability not merely to be mobile sensors and actuating devices but to serve as long-term carriers of local knowledge that transform chaotic events into structured semantic maps and distributed policies of behaviour. They combine low latency, operational resilience, and resource savings, creating a foundation for scalable, secure, and interoperable transport systems suitable for operation in real and even adversarial conditions.

As a result, the integration of blockchain into inter-agent protocols for loitering models transforms the AIV system into a self-regulating environment (Fig. 6) that is capable of adaptively changing according to the situation and current needs.

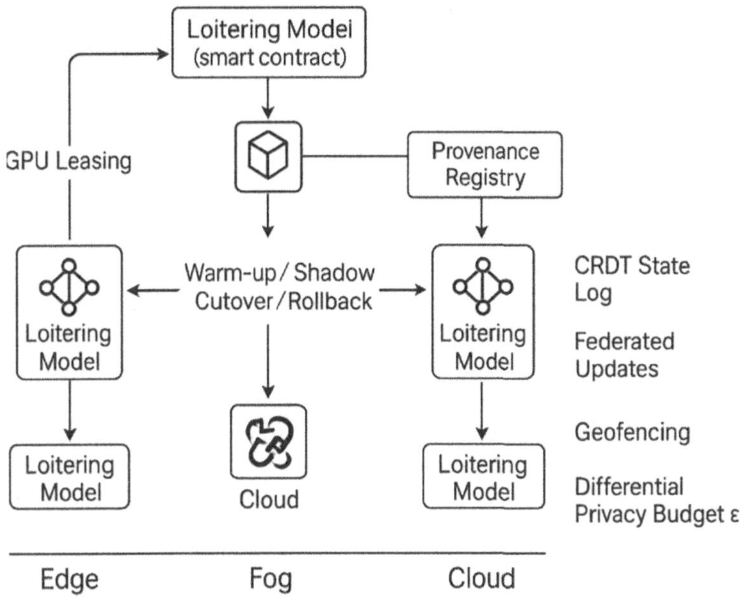

Fig. 6 Diagram of the relationships between the blockchain ledger, resource auctions, smart contracts, and agents (the image was generated using the OpenAI gpt-image-1 model)

11 Conclusions

Summarizing the foregoing, it should be noted that when the blockchain infrastructure is laid into the very fabric of a multi-agent AIV system, such systems reach a qualitatively new level of autonomy. In this case, the blockchain serves as an internal source of objectivity and an interdomain bus for decision coordination, rather than yet another service layered atop control processes. The AIV's own onboard log orders the events of sensor, planning, and executive AI agents into a single timeline, provides verifiability without dependence on external connectivity, and forms minimal yet meaningful artefacts for subsequent registration in domain registries. External permissioned hashing chains with BFT-class consensus accept only compact digests in the form of Merkle roots or zero-knowledge proofs, separating local performance from the interdomain layer of trust. The real-time control loop remains isolated from slow notarization, which preserves millisecond latency budgets, and micro-batching together with hardware-accelerated hashing does not add jitter to critical control loops. Such a decomposition yields transparent auditability, resilience to faults and failures, and the possibility of offline operation with subsequent reconciliation of history by the rule of the largest finalized prefix.

The hierarchical organization of urban registries resolves the tension between scale and privacy. Domain chains for transport, aviation, energy, and security operate with their own policies and finalization speeds, whereas the common urban trust layer merely aggregates facts, arbitrates disputes, and does not block local processes. Formally verified bridges, a unified event model, and hybrid cryptography with post-quantum algorithms make such a system governable and durable over the long term. For mobile objects, predictable procedures are supported for migration between domains and settlements with temporary passes, pre-issued authorities and resource budgets, as well as policies for safe offline modes. The same principles work uniformly for ground robotic vehicles, swarms of drones, flotillas of USVs, and mixed UAV + USV + UUV configurations, where dual affiliation and session thinking ensure continuity of history without duplication of large data.

A separate vector of development concerns loitering agents. Their spatiotemporal presence near data sources reduces latency, equalises load across nodes, increases robustness to link disruptions, and converts episodic events into long-lived semantic maps. Portability of weights and state together with a containerised microservice environment, lightweight context adapters, and local migration rules with checkpoints make it possible to flexibly move intelligence vertically across Edge, Fog, and Cloud and laterally between peer nodes without loss of controllability or security. As a result AIVs become participants in a distributed economy of computation and data. Their resources are allocated transparently through auctions, interactions are governed by smart contracts, and learning becomes collective and verifiable.

The key conclusion is that to obtain scalable, interoperable, and secure AIVs, one should design not their individual modules but an integral ecosystem based on three fundamental components. Two of them are traditional for blockchain systems, namely architectural separation of motion control processes from accounting and trust, and the creation of a hierarchy of registries with clear boundaries of responsibility, formally defined policies, and lightweight inter-network proofs. The third component is mobile intelligence in the form of loitering agents that operate alongside the data and carry knowledge together with their own state. Such a triad ensures the required balance between performance, transparency, privacy, and fault tolerance, opening the way to effective operation of autonomous transport in complex urban and non-urban environments.

References

1. Zhou, Z.-H.: Ensemble Methods: Foundations and Algorithms (Chapman & Hall/CRC Machine Learning & Pattern Recognition). CRC Press, Taylor & Francis Group, Boca Raton, FL, USA (2012)
2. Chen, Z., Deng, Y., Wu, Y., Gu, Q., Li, Y.: Towards Understanding Mixture of Experts in Deep Learning. ArXiv 2208.02813 [cs.LG] (2022). https://doi.org/10.48550/arXiv.2208.02813
3. Mistral AI team: Mixtral of experts: A high quality Sparse Mixture-of-Experts (2023). https://mistral.ai/news/mixtral-of-experts/
4. Vakulenko, M., Slyusar, V.: Automatic smart subword segmentation for the reverse Ukrainian physical dictionary task. CEUR Work. Proc. **3723**, 59–73 (2024)
5. Durante, Z., Huang, Q., Wake, N., Gong, R., Park, J.S., Sarkar B., Taori, R., Noda, Y., Terzopoulos, D., Choi, Y., Ikeuchi, K., Vo, H., Fei-Fei, L.: Agent AI: Surveying the Horizons of Multimodal Interaction (2024). https://arxiv.org/abs/2401.03568

6. Slyusar, V.: Distributed multi-agent systems based on the mixture of experts architecture in the context of 6G wireless technologies. In: Dovgyi, S., et al. (eds.) Applied Innovations in Information and Communication Technology. ICAIIT 2024. Lecture Notes in Networks and Systems, vol. 1338, pp. 81–110. Springer (2025). https://doi.org/10.1007/978-3-031-89296-7_6
7. Baier, T., Baez Santamaria, S., Vossen, P.: A modular architecture for creating multimodal agents (2022). https://arxiv.org/abs/2206.00636
8. Slyusar, V.I., Kondratenko, Y.P., Shevchenko, A.I., Yeroshenko, T.V.: Some aspects of artificial intelligence development strategy for mobile technologies. J. Mob. Multimed. 20(3), 525–554 (2024). https://doi.org/10.13052/jmm1550-4646.2031
9. Haber, S., Stornetta, W.S.: How to time-stamp a digital document. J. Cryptol. **3**, 99–111 (1991). https://doi.org/10.1007/BF00196791
10. Merkle, R.C.: A digital signature based on a conventional encryption function. In: Advances in Cryptology CRYPTO '87, LNCS 293, Santa Barbara, CA, USA, pp. 369 378 (1987). https://doi.org/10.1007/3-540-48184-2_32
11. Nakamoto, S.: Bitcoin: A Peer-to-Peer Electronic Cash System (2008). https://bitcoin.org/bitcoin.pdf. Accessed 1 Oct 2025
12. Ota, D.: Towards verification of NATO generic vehicle architecture-based systems. In: Proceedings of the Twenty-First International Command and Control Research and Technology Symposium (21ST ICCRTS), London, UK. p. 22 (2016). https://www.researchgate.net/publication/309027098_Towards_Verification_of_NATO_Generic_Vehicle_Architecture-Based_Systems
13. Bhardwaj, D., Beniwal, A., Chaudhari, S., Kalyan, A., Rajpurohit, T., Narasimhan, K. R., Deshpande, A., Murahari, V.: Agent context protocols enhance collective inference. https://arxiv.org/abs/2505.14569 (2025)
14. Habler, I., Huang, K., Narajala, V. S., & Kulkarni, P. (2025). Building a secure agentic AI application leveraging A2A protocol. arXiv. https://arxiv.org/abs/2504.16902
15. Jeong, C.: A study on the MCP × A2A framework for enhancing interoperability of LLM-based autonomous agents (2025). https://arxiv.org/abs/2506.01804
16. Liu, J., Yu, K., Chen, K., Li, K., Qian, Y., Guo, X., Song, H., Li, Y.: ACPs: Agent collaboration protocols for the Internet of Agents (2025). https://arxiv.org/abs/2505.13523
17. Martineau, K.: The simplest protocol for AI agents to work together. IBM Research (2025). https://research.ibm.com/blog/agent-communication-protocol-ai
18. 6G Flagship: 6G White Papers. 6G Flagship (2023). https://www.6gflagship.com/white-papers
19. 6G Flagship: Books on 6G. 6G Flagship (2023). https://www.6gflagship.com/books
20. Kong, L., Tan, J., Huang, J., Chen, G., Wang, S.: Edge-computing-driven internet of things: a survey. ACM Comput. Surv. **55**(8), 174 (2022). https://huangjunqin.com/assets/pdf/KongCSUR2022Edge.pdf
21. Castro, M., Liskov, B.: Practical byzantine fault tolerance. In: Proceedings of 3rd Symposium on Operating Systems Design and Implementation (OSDI '99), New Orleans, LA, USA, pp. 173–186 (1999). https://www.usenix.org/conference/osdi-99/practical-byzantine-fault-tolerance. Accessed 2 Oct 2025
22. Slyusar, V., Kozlov, V., Pochernin, S., Nalapko, I.: Conceptual foundations of the swarm employment of unmanned aerial vehicles as intelligent means of electronic warfare. Technol. Audit. Prod. Reserv. **3/2**(83), 71–80 (2025). https://doi.org/10.15587/2706-5448.2025.329989
23. Wei, W., Wang, J., Fang, Z., Chen, J., Ren, Y., Dong, Y.: 3U: joint design of UAV-USV-UUV networks for cooperative target hunting. IEEE Trans. Veh. Technol. **72**(3), 4085–4090 (2023). https://doi.org/10.1109/TVT.2022.3220856
24. Potter, J., Alves, J., Green, D., Zappa, G., Nissen, I., McCoy, K.: The JANUS underwater communications standard. In: 2014 Underwater Communications and Networking (UComms), Sestri Levante, Italy, pp. 1–4 (2014). https://doi.org/10.1109/UComms.2014.7017134
25. JANUS creates a new era for digital underwater communications. Robohub. https://robohub.org/janus-creates-a-new-era-for-digital-underwater-communications/

26. Slyusar, V., Bihun, N.: Innovations in marine unmanned vehicles: image quality enhancement with segmented autoencoders. In: 2024 IEEE 7th International Conference on Actual Problems of Unmanned Aerial Vehicles Development, APUAVD 2024—Proceedings, pp. 209–214 (2024). https://doi.org/10.1109/APUAVD64488.2024.10765872
27. Bonomi, F., Milito, R., Zhu, J., Addepalli, S.: Fog computing and its role in the internet of things. In: Proceedings of 1st ACM MCC Workshop on Mobile Cloud Computing, Helsinki, Finland, pp. 13–16 (2012). https://conferences.sigcomm.org/sigcomm/2012/paper/mcc/p13.pdf
28. Slyusar, V.I., Smolyar, V.G.: Non-orthogonal discrete frequency signal modulation method for narrowband communication channels. Izv. Vyss. Uchebnykh Zavedenij. Radioelektron. **47**(4), 53–59 (2004)
29. Slyusar, V.I., Smolyar, V.G.: Communication channels frequency multiplexing on the basis of superrayleigh signals resolution *Izvestiya Vysshikh Uchebnykh Zavedenij.* Radioelektronika **46**(7), 30–39 (2003)
30. Sliusar, I., Voloshko, S., Smolyar, V., Slyusar, V.: Next generation optical access based on N-OFDM with decimation. In: 2016 3rd International Scientific-Practical Conference Problems of Infocommunications Science and Technology, PIC S and T 2016—Proceedings, pp. 192–194 (2017)
31. Hadani, R., Rakib, S., Tsatsanis, M., Monk, A., Goldsmith, C., Molisch, A.F., Calderbank, R., Loyka, S.: Orthogonal time frequency space modulation. In: Proceedings of IEEE Wireless Communications and Networking Conference (WCNC), pp. 1–6 (2017). https://doi.org/10.1109/WCNC.2017.7925924
32. Liang, Y., Li, L., Fan, P., Guan, Y.: Doppler resilient orthogonal time frequency space (OTFS) systems based on index modulation. In: Proceedings of IEEE VTC'2020-Spring, pp. 1–5 (2020)
33. Son, S., Lee, J., Park, Y., Park, Y., Das, A.K.: Design of blockchain-based lightweight V2I handover authentication protocol for VANET. IEEE Trans. Netw. Sci. Eng. **9**(3), 1346–1358 (2022). https://doi.org/10.1109/TNSE.2022.3142287
34. Wang, S., Wu, Y., Wen, K., Zhou, X., Hu, B., Xie, Q.: An improved blockchain-based lightweight vehicle-to-infrastructure handover authentication protocol for vehicular Ad hoc networks. Mathematics **13**(7), 1118 (2025). https://doi.org/10.3390/math13071118

Blockchain-Augmented AI Empowered Autonomous UAV Security Systems: Towards Trustworthy Airspace Operations on the Internet of Drones (IoD)

Mohtasin Golam, Rubina Akter, Esmot Ara Tuli, and Dong-Seong Kim

Abstract The rapid adoption of unmanned aerial vehicles (UAVs) across civilian, commercial, and defense sectors has intensified concerns over the security and integrity of Internet of Drones (IoD) networks. This proliferation has also raised serious concerns regarding the security and privacy of UAV communications, particularly in light of emerging cyber threat landscapes. This chapter proposes exploring secure, intelligent architecture for the Internet of Drones (IoD), drawing on blockchain-based deep learning (DL) integrated synergistic frameworks for the development of real-time unauthorized UAV localization. Specifically, it will explore how emerging frameworks combine resilient blockchain consensus (e.g., PoA, DPoA) with high-performing angle-of-arrival (AoA) estimation via lightweight CNN architectures. These systems demonstrate scalable, zone-based authentication and autonomous threat identification, establishing a blueprint for resilient drone communication management in smart airspaces. Additionally, it will investigate the role of blockchain in secure UAV identity verification, navigation authorization, and tamper-resistant communication, serving as cornerstones for a zero-trust UAV ecosystem. The chapters conclude by exploring future enhancements such as federated learning for adaptive model refinement and cross-domain consensus schemes to advance a zero-trust, scalable, and trustworthy IoD ecosystem.

M. Golam · E. A. Tuli
ICT Convergence Research Center, Kumoh National Institute of Technology, Gumi, South Korea
e-mail: gmoh248@kumoh.ac.kr

E. A. Tuli
e-mail: tuli.cse.bd@gmail.com

R. Akter
Department of Electrical and Software Engineering, University of Calgary, Calgary, Canada
e-mail: rubina.akter@ucalgary.ca

D.-S. Kim (✉)
NSLab Inc., ICT Convergence Research Center, Gumi, South Korea
e-mail: dskim@kumoh.ac.kr

S. O. Ajakwe and D.-S. Kim (eds.), *Blockchain for Security of Autonomous Intelligent Vehicles*, Studies in Big Data 194,
https://doi.org/10.1007/978-3-032-20851-4_5

Keywords Authentication · Blockchain · Convolutional neural network (CNN) · Deep learning (DL) · Internet of drones (IoD) · UAV security · Zero trust

1 Introduction

The Internet of Drones (IoD) represents a paradigm shift in aerial systems, transforming isolated unmanned aerial vehicles (UAVs) into a cohesive, service-oriented cyber-physical ecosystem. At the core, UAVs operate as both networked sensors and actuators, exchanging telemetry data—such as positional data, altitude, longitude, latitude, speed, velocity, and environmental measurements and receiving mission instructions through resilient air-to-ground (A2G) and air-to-air (A2A) communication links [1]. These links leverage software-defined radios, mesh networking protocols, and emerging 5G/6G air interfaces to sustain low-latency, high-throughput connections under dynamic flight conditions [1]. Furthermore, through these resilient links, UAVs exchange real-time data and control messages, enabling dynamic reconfiguration, cooperative behaviors, and robust fault mitigation [2].

Above the physical connectivity layer, the IoD provides high-level services that abstract UAV capabilities into modular functions. Collision avoidance operates as a distributed safety service in which each UAV uses on-board sensors—radar, LiDAR, and vision systems to detect proximate obstacles, while edge computing nodes compute deconfliction trajectories and broadcast corrective advisories [3]. Geofencing services enforce spatial constraints by validating flight plans against preconfigured polygonal zones or dynamically issued no-fly directives, triggering automated rerouting or mission aborts as needed [4]. Payload handoff services facilitate mid-air transfers of sensor data or physical packages: reconnaissance drones may stream high-resolution imagery to relay UAVs, which forward aggregated data to ground control stations along optimized routes [5].

The orchestration layer of the IoD converges mission planning, dynamic task allocation, and cooperative swarm coordination [5]. Centralized control centers effectively break down important tasks, such as area surveillance, search and rescue (SAR), and precision agriculture mapping, into manageable subtasks. By assigning these subtasks to individual UAVs based on their current status, energy levels, and sensor configurations, they enhance operational efficiency and ensure that each mission is carried out successfully [6]. Dynamic reallocation protocols adjust task assignments in response to environmental changes or emergent mission priorities, ensuring resilience to UAV failures or communication disruptions. In fully decentralized deployments, consensus algorithms enable peer-to-peer negotiation of task roles and flight paths, enhancing scalability and eliminating single points of failure (SPoF) [7].

Despite these technological advancements, the IoD remains vulnerable to a range of cyber-physical security threats. Many UAVs transmit control and telemetry data over unencrypted channels, rendering them susceptible to eavesdropping, message injection, and unauthorized command execution [2]. Adversaries can take control of drones or manipulate mission-critical data by exploiting weak security systems. Additionally, the video and images transmitted by these drones can be altered during transmission, leading to data tampering and compromising confidentiality during urgent operations where accurate information is crucial [8]. Regulatory rules like Remote ID require drones to broadcast their location and the identity of their operators. This can unintentionally reveal sensitive information, such as flight paths, mission plans, and personnel details, to individuals with bad intentions [2].

These vulnerabilities have been extensively documented in recent surveys of UAV communication networks and flying ad hoc networks (FANETs), which highlight attack vectors such as GPS spoofing, signal jamming, replay attacks, and Sybil node insertion [8–10]. The absence of robust intrusion detection mechanisms further amplifies risk, as compromised UAVs may participate in swarm operations, propagating false data or launching coordinated denial-of-service (DoS) attacks [11]. Traditional centralized security architectures face significant scalability challenges; as the number of UAVs expands to hundreds or even thousands, latency and bandwidth limitations hinder effective real-time authentication and anomaly detection responses [7].

To address these challenges, blockchain technology has emerged as a compelling foundation for securing IoD networks. By virtue of its decentralized ledger, cryptographic integrity, and programmable smart contracts, blockchain enables multiple critical security functions without reliance on a single trusted authority [12]. A permissioned blockchain can record details about drones, including their identities, cryptographic keys, and flight permissions. This creates a secure, immutable flight registry that helps prevent unauthorized access and resolve conflicts in shared airspace [12]. Each UAV's public key and registration details could be stored on-chain, while private credentials remain confined to trusted distributed ledgers, thwarting identity spoofing and cyber attacks.

Smart contracts deployed on the blockchain further enhance security by codifying geofencing and operational policies, thereby ensuring a higher level of security [13]. Predefined rules, no-fly zones restrictions, altitude limits, and time-restricted flight could be encoded into contract logic [14]. When a UAV proposes a flight plan or emits a GPS update, the smart contract automatically validates the coordinates against policy constraints and returns an authorization verdict. In the event of a violation, the contract triggers an immediate reroute instruction, issues an alert to neighboring UAVs, and logs the incident for post-mission forensic analysis [15]. This automated enforcement reduces human oversight errors and ensures consistent policy application across heterogeneous IoD networks.

Thus, this chapter addresses how the fusion of blockchain and artificial intelligence (AI) can secure the IoD by answering three core questions: (i) How can decentralized ledgers and smart contracts authenticate UAV identities and enforce flight policies? (ii) How can AI models detect and localize unauthorized drones in real time? iii) and how can their integration create a zero-trust, transparent framework that thwarts spoofing, tampering, and unauthorized access without introducing prohibitive latency or resource overhead? To explore these questions, we explore how to create an architecture that combines a permissioned blockchain, ensuring immutable flight registries, geofencing enforcement, and automated policy execution [16], with edge-deployed convolutional neural networks (CNNs) that analyze angle-of-arrival (AoA) signals for the detection of intruding UAVs [12]. This further demonstrates how smart contracts can automatically validate flight plans, revoke permissions upon policy violations, and record security events in tamper-proof audit trails, while AI modules provide rapid, on-the-fly threat assessments [15]. Together, this convergence could deliver a robust, scalable IoD security solution that maintains continuous trust verification, supports real-time decision making, and provides transparent auditability without compromising UAV performance.

2 Blockchain-Based Access Control and Trust Management Framework

The foundation of secure Internet of Drones (IoD) operations lies in establishing a robust, decentralized access-control mechanism that authenticates UAV identities while preventing unauthorized access. This section presents a comprehensive blockchain-based trust management framework that eliminates single points of failure (SPoF) inherent in centralized authentication systems, providing tamper-resistant audit trails for all UAV operations.

2.1 Hierarchical Blockchain Architecture

The proposed framework utilizes a hierarchical blockchain structure comprising two distinct layers: a private blockchain deployed within individual clusters to maintain confidentiality between zones, and a consortium blockchain that connects multiple clusters to the central control server (CCS). This dual-layer approach ensures that sensitive operational data remains distinguished while enabling cross-cluster coordination when necessary [12].

Each zone within the IoD network is managed by a Zone Edge Server (ZES), which serves as a validator node in the blockchain. The ZES maintains three core components: a distributed ledger for storing UAV credentials and flight authorizations, an authentication unit for processing access requests, and a control unit for managing cryptographic key distribution [16]. This hierarchical design addresses scalability concerns by distributing computational load across multiple edge nodes rather than relying on a single centralized authority, as illustrated in Fig. 1 [12].

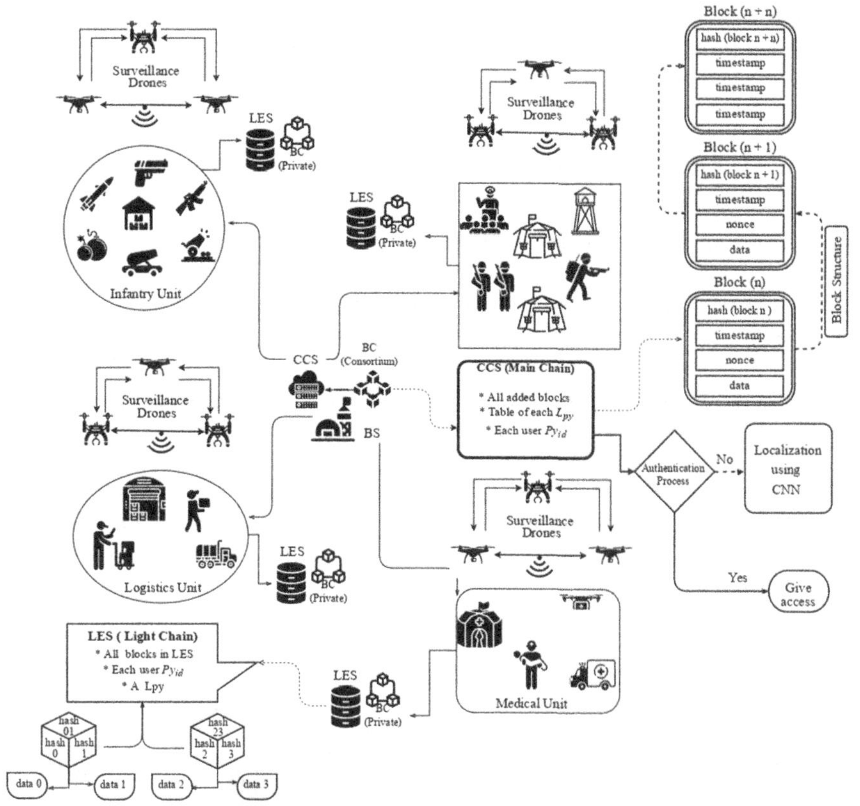

Fig. 1 Hierarchical blockchain architecture for IoD network showing the integrated blockchain-based UAV authentication and CNN-based localization framework [12]

2.2 *Permissioned Blockchain and Consensus Mechanism*

The framework utilizes a permissioned blockchain network with a customized Proof-of-Authority (PoA) consensus algorithm, specifically designed for resource-constrained UAV environments. Unlike traditional Proof-of-Work (PoW) mechanisms that require extensive computational resources, the proposed Drone-based PoA (DPoA) consensus ensures energy-efficient transaction validation while maintaining security against Byzantine faults [16, 17].

The DPoA mechanism operates through pre-authorized validator nodes, typically high-capability military assets such as transporter erector launchers (TELs), tanks, or ground control stations, which possess sufficient computational power to participate in the validation process [17]. This approach eliminates the energy-intensive mining requirements of DPoA while preserving the security necessary for mission-critical operations. Experimental results demonstrate that DPoA achieves consistent block mining times of approximately one second, compared to variable processing times in PoA and PoW systems, which depend on the difficulty levels.

2.3 UAV Authentication and Registration Process

Figure 2 illustrates the authentication framework, which begins with UAV registration, during which each drone submits encrypted credentials, including its longitude, latitude, MAC address, and a cryptographic hash, to the local ZES [12]. The registration process generates a pseudonymous identifier (PyID) for each UAV, which is stored in both the ZES pseudo-list and the CCS blockchain ledger. This design enables UAVs to migrate between zones without requiring re-authentication, significantly reducing network latency and computational overhead.

When a UAV requests network access, the ZES authentication unit first queries its local pseudo-list for the drone's PyID. If the identifier is found locally, authentication is granted immediately [17]. Otherwise, the system queries the CCS blockchain to verify if the UAV belongs to a different zone. UAVs that fail both authentication checks are flagged as potential intruders, with their detection triggering the AI-based localization subsystem described in subsequent sections [12].

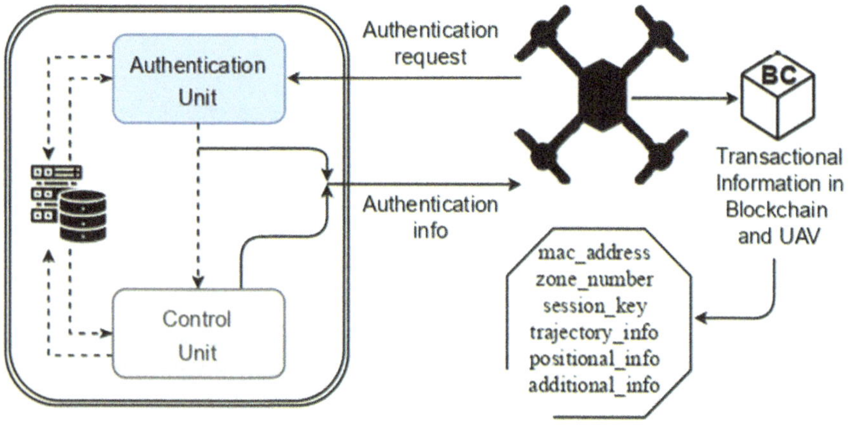

Fig. 2 Blockchain transaction structure for IoD authentication showing the data flow between UAV authentication components and blockchain storage [16]

2.4 Smart Contract-Based Geofencing and Policy Enforcement

Smart contracts deployed on the blockchain automate the enforcement of geofencing policies and flight restrictions, ensuring compliance with regulations. These contracts encode no-fly zones, altitude limits, and temporal restrictions as immutable rules that are automatically validated against every UAV position update [12]. When a drone's proposed flight path violates geospatial constraints, the smart contract immediately revokes flight authorization, issues automated rerouting commands, and logs the violation for forensic analysis.

The smart contract architecture supports dynamic policy updates through consensus-based governance mechanisms. Authorized personnel can propose policy modifications such as temporary no-fly zones for special operations, which are validated by the DPoA consensus and automatically propagated across all network nodes [16]. This capability ensures rapid adaptation to changing operational requirements while maintaining audit trails of all policy changes.

2.5 Blockchain-Enabled Traffic Management and Airspace Coordination

2.5.1 Congestion Management Through Decentralized Coordination

As unmanned aerial vehicle operations scale toward thousands of simultaneous flights in dense urban airspace, centralized traffic management architectures face fundamental bottlenecks in processing real-time coordination requests and maintaining situational awareness across heterogeneous operators [18]. The proposed blockchain framework addresses these congestion challenges through distributed consensus mechanisms that enable autonomous conflict resolution among UAVs without requiring continuous central authority intervention [19]. Each participating UAV broadcasts its intended four-dimensional trajectory as a blockchain transaction that includes spatial coordinates, altitude bands, velocity vectors, and temporal windows. Smart contracts deployed on the blockchain network automatically detect spatiotemporal conflicts by computing trajectory intersections and validating separation minima against other registered flight plans in the distributed ledger. When conflicts are identified, the smart contract executes priority-based resolution algorithms that systematically evaluate mission criticality, remaining battery capacity, payload time-sensitivity, and regulatory constraints to determine optimal deconfliction strategies. Emergency response UAVs carrying time-critical medical supplies or conducting search-and-rescue operations receive the highest priority classification, followed by commercial delivery drones operating under service-level agreements, and finally recreational operations with flexible scheduling tolerance [20].

The blockchain maintains a real-time occupancy map of airspace segments by aggregating all active flight plans and dynamically updating availability as UAVs complete their routes or modify trajectories. This distributed airspace state enables UAVs to autonomously query available corridors and dynamically reroute around congested zones without centralized traffic flow management intervention [18]. The decentralized coordination approach significantly reduces decision-making latency by eliminating round-trip communication delays to central servers and removing single points of failure that plague traditional centralized air traffic control systems. Performance evaluations of blockchain-based unmanned traffic management systems demonstrate transaction processing capabilities exceeding 1,000 flight plan validations per second using permissioned consensus protocols optimized for UAV networks [21]. To manage high-frequency telemetry data that would overwhelm on-chain storage, the architecture employs a hybrid on-chain and off-chain data management architecture, where detailed trajectory waypoints and sensor telemetry are stored in distributed file systems such as the InterPlanetary File System (IPFS), while cryptographic hashes and clearance decisions are anchored on the immutable blockchain for auditability [15]. This design pattern balances transparency requirements with scalability constraints, enabling the system to support dense urban airspace operations involving hundreds of concurrent UAVs per cubic kilometer.

2.5.2 Dynamic Airspace Sharing and Corridor Allocation

The framework implements dynamic airspace sharing through tokenized corridor allocation, where three-dimensional flight corridors are represented as digital assets on the blockchain that can be reserved, traded, and reallocated based on operational demand. Urban airspace is partitioned into volumetric cells defined by latitude-longitude boundaries, altitude layers, and temporal availability windows that collectively form a discrete airspace reservation grid. UAVs request temporary exclusive or shared access rights to specific corridor segments by initiating smart contract transactions that specify entry time, duration, intended trajectory, and operational parameters. The smart contract evaluates corridor availability by querying the blockchain ledger for the current reservation state and validating the requesting UAV's credentials, including registration status, insurance coverage, and operator licensing. For corridors with sufficient separation capacity, the smart contract grants shared usage rights to multiple UAVs whose trajectories maintain minimum separation distances as defined by regulatory safety standards, which are automatically enforced by geometric conflict-detection algorithms.

When high-priority missions such as emergency medical transport or law enforcement operations require exclusive corridor access, the smart contract automatically notifies affected UAVs holding shared reservations in the target corridor and triggers coordinated rerouting protocols. The notification messages contain alternative corridor suggestions computed by pathfinding algorithms that minimize detour distance while respecting remaining battery constraints. Expired corridor tokens are automatically revoked by smart contract logic at the end of the reservation time window, ensuring that airspace resources are recycled efficiently and preventing unauthorized lingering in restricted zones. This tokenized approach provides transparent, auditable corridor allocation while enabling flexible capacity management during peak demand periods via dynamic pricing mechanisms encoded in smart contracts. Studies of blockchain-based airspace reservation systems demonstrate that token-based allocation reduces corridor waste by 35% compared to static time-slot scheduling and enables 60% higher airspace utilization during asymmetric demand patterns. The system supports secondary markets where operators can transfer unused corridor reservations to other UAVs through blockchain transactions, creating economic incentives for efficient resource utilization and enabling flexible response to operational disruptions.

2.5.3 UAV Prioritization Using Reputation-Based Consensus

The system incorporates a multidimensional reputation scoring mechanism in which UAVs accumulate trust metrics based on historical compliance with flight regulations, successful mission completion rates, adherence to traffic protocols, and safety incident records. These reputation scores are immutably recorded on the blockchain through signed transactions that aggregate validator attestations, preventing unilateral score manipulation by individual participants. The reputation framework evaluates multiple behavioral dimensions, including trajectory prediction accuracy (how closely actual flight paths match submitted flight plans), response time to traffic advisories issued by the coordination system, and compliance with dynamic airspace restrictions during temporary flight restrictions. Each dimension is weighted by its impact on overall system safety and efficiency, with serious violations, such as geofence breaches or near-miss incidents, resulting in substantial reputation penalties that persist in the permanent ledger.

During airspace contention scenarios in which multiple UAVs request access to limited corridor capacity, smart contracts use reputation scores as tie-breaking criteria to prioritize UAVs with demonstrated safety records. Emergency service UAVs operated by verified first-responder organizations maintain consistently high reputation scores through cryptographically signed credential verification, ensuring unimpeded access during critical operations regardless of network congestion. Commercial operators can systematically improve their reputation by consistently complying with regulations and participating in cooperative traffic management protocols, earning preferential treatment in corridor allocation and reduced reservation fees. The blockchain-based reputation system prevents gaming by cryptographic verification of compliance data sourced from multiple independent validators, including ground-based surveillance systems, satellite observation networks, and peer UAV witnesses, which collectively validate reported behavior.

Recent implementations of distributed reputation management for drone networks demonstrate that reputation-weighted prioritization reduces average corridor waiting times by 40% while maintaining fairness through transparent score computation and appeal mechanisms [19]. The system creates economic incentives for responsible

UAV operation by translating reputation scores into tangible operational benefits, while maintaining equitable access for new operators establishing operational track records through provisional scoring that emphasizes recent behavior over historical gaps. Reputation decay mechanisms gradually reduce the influence of outdated historical data, ensuring that rehabilitated operators can recover from past incidents through sustained compliant operations [22].

2.5.4 Scalability and Real-Time Performance Optimization

To support air-traffic-scale operations involving thousands of concurrent UAVs in metropolitan airspace, the blockchain architecture employs hierarchical sharding strategies that partition the coordination workload across multiple parallel consensus domains. The airspace is divided into geographic zones, each managed by a dedicated blockchain shard that processes local traffic coordination transactions with minimal cross-shard dependencies. Intra-zone flights that remain within a single geographic partition are validated entirely within the local shard, achieving transaction finality in sub-second timeframes without global consensus overhead. Cross-zone flights that traverse multiple geographic partitions require atomic transactions coordinated through inter-shard communication protocols that employ two-phase commit semantics to ensure consistent state across zone boundaries.

Layer-two scaling solutions, including state channels and plasma chains, enable high-frequency position updates between cooperating UAVs without burdening the main blockchain with excessive transaction volume. Pairs or groups of UAVs engaged in coordinated operations, such as swarm maneuvers, establish temporary off-chain state channels where position synchronization and cooperative control messages are exchanged with microsecond-scale latency. Periodic settlement transactions aggregate off-chain state updates and anchor cryptographic commitments to the immutable blockchain ledger, providing verifiable audit trails while maintaining real-time responsiveness. Performance evaluations using Hyperledger Fabric permissioned blockchain implementations demonstrate transaction throughput exceeding 10,000 operations per second with average latency below 200 ms, sufficient to support dense urban airspace operations while maintaining sub-second response for critical safety functions.

The consensus mechanism employs practical Byzantine fault tolerance variants optimized for UAV networks, where validator nodes are distributed across ground infrastructure, cellular base stations, and airborne relay platforms to ensure resilience against localized failures. Permissioned consensus reduces the computational overhead compared to proof-of-work systems while maintaining strong consistency guarantees appropriate for safety-critical aviation applications. Studies of blockchain integration in UAV networks confirm that permissioned architectures achieve a three-order-of-magnitude improvement in energy efficiency compared to public blockchain alternatives, making them feasible for resource-constrained edge computing environments typical of unmanned traffic management infrastructure.

3 AI-Driven Intruder UAV Detection

The proliferation of sophisticated threats in drone networks necessitates advanced detection mechanisms that can identify unauthorized UAVs in real-time while operating under resource constraints. This section presents comprehensive AI-driven approaches for intruder UAV detection, emphasizing lightweight convolutional neural networks (CNNs) optimized for angle-of-arrival (AoA) estimation and direction-of-arrival (DoA) localization in Internet of Drones (IoD) environments [12, 17, 23].

3.1 Signal Model and Antenna Array Configuration

The foundation of AI-driven UAV detection relies on sophisticated signal processing techniques that capture and analyze radio-frequency (RF) emissions from aerial vehicles, as shown in Fig. 3. The proposed system employs a non-uniform linear antenna array (NULA) configuration specifically designed to enhance spatial resolution and detection accuracy. The NULA configuration utilizes antenna element spacings of $\lambda/2$, providing superior aperture characteristics compared to uniform linear arrays. When an unauthorized UAV enters the monitored airspace, independent electromagnetic waves are received by the $\rho - antenna$ elements array. The signal preprocessing involves analog-to-digital conversion, followed by in-phase and quadrature (IQ) sampling, which creates high-dimensional data matrices suitable for deep learning analysis.

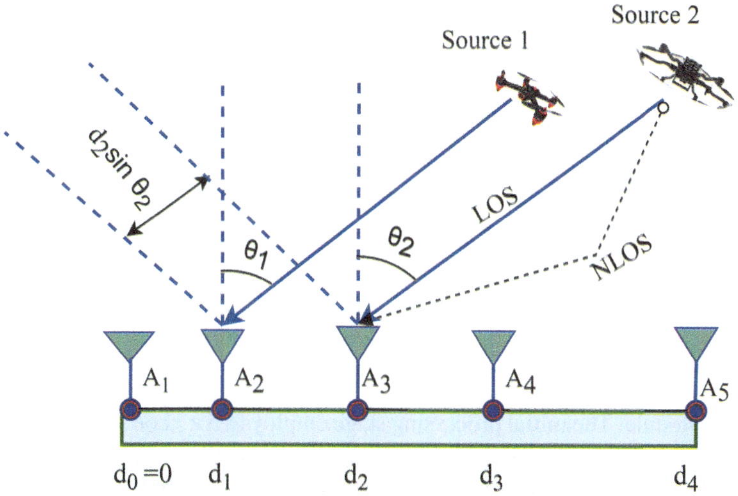

Fig. 3 Signal processing model to estimate the direction/angle of arrival [12]

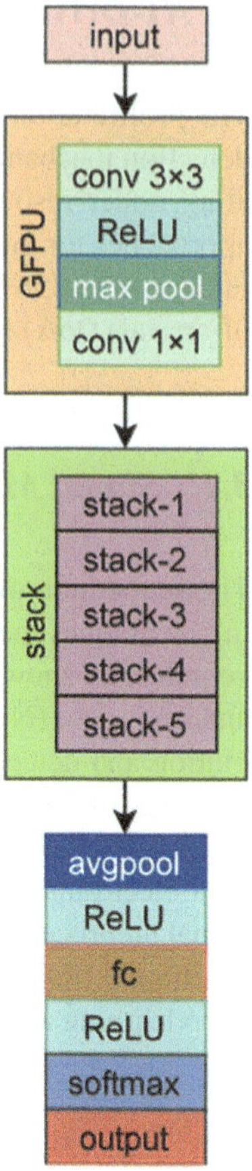

Fig. 4 Network architecture of LRCANet for AOA estimation: overall structure of LRCANet

3.2 Lightweight CNN Architectures for UAV Localization

3.2.1 LRCANet Architecture

The Lightweight Residual Convolution Architecture Network (LRCANet) represents a sophisticated approach to AoA estimation, specifically designed for IoD applications. The architecture comprises two primary modules: the Generic Feature Processing Unit (GFPU) and the Stack modules, organized using asymmetric convolutional layers and residual connectivity (Figs. 4 and 5).

- GFPU Module: The initial processing stage employs a 3×3 convolution layer with stride (1, 1) followed by ReLU activation and max pooling.
- Stack Modules: Five cascaded stack modules process features using asymmetric convolution layers (1×3 and 3×1), offering greater learning efficiency than traditional 3×3 kernels while maintaining reduced trainable parameters. Each stack

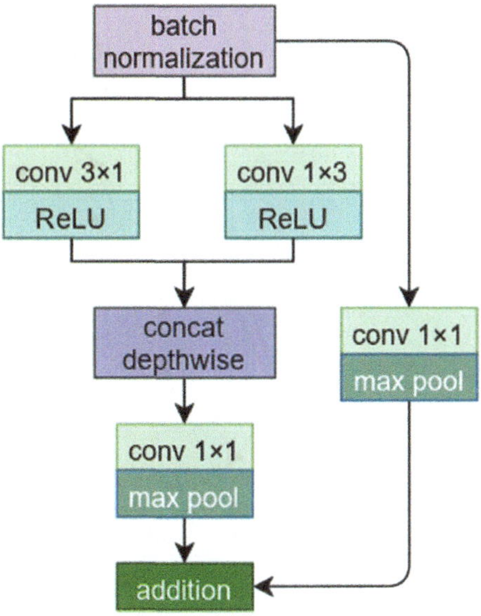

Fig. 5 Network architecture of LRCANet for AOA estimation: stack configurations

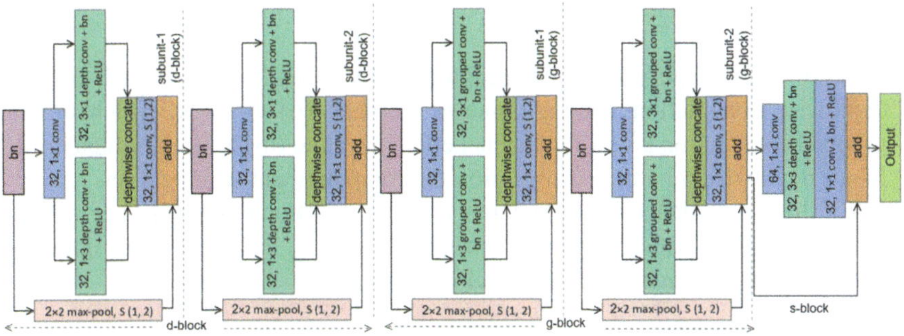

Fig. 6 Illustration of the RAT module with cascaded connections of two d-blocks, a g-block, and one s-block [12]

incorporates batch normalization, depth-wise concatenation, and residual connections to mitigate vanishing gradient problems.

3.2.2 IoMT-Net Architecture

The IoMT-Net framework introduces a more sophisticated dual-module approach, comprising the Resource Accuracy Tradeoff (RAT) and Unique Feature Extraction and Accuracy Boosting (UAB) modules. This architecture achieves superior performance, with 97.63% accuracy at 10 dB SNR (Fig. 6).

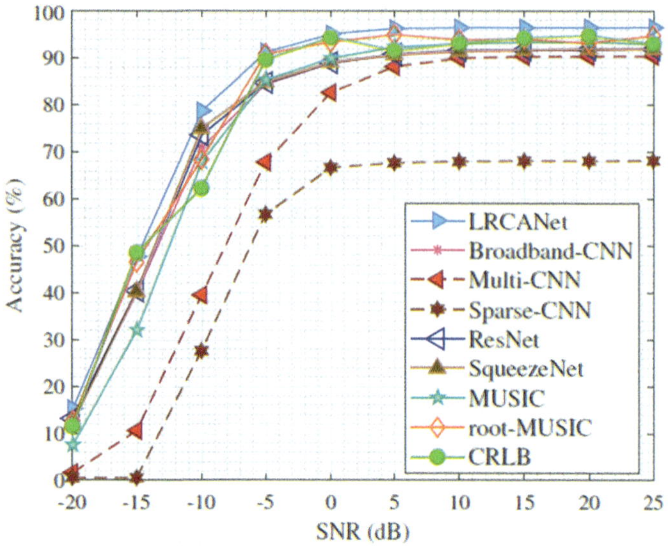

Fig. 7 Method comparison in terms of classification accuracy [17]

- RAT Module: It integrates three specialized blocks: (i) D-block, which employs depthwise convolution with asymmetric kernels; (ii) G-block, which utilizes grouped convolution layers for channel efficiency; and (iii) S-block, which implements depthwise separable convolution for computational optimization.
- UAB Module: Enhances discriminative feature extraction through deeper network structures with periodic depth dimension expansion, contributing to improved accuracy while maintaining computational efficiency.

3.3 Performance Analysis and Computational Efficiency

Comprehensive testing across signal-to-noise ratio (SNR) levels from -20dB to +25dB reveals robust performance characteristics. At challenging SNR conditions (-10dB), LRCANet maintains 89% accuracy while IoMT-Net achieves 92% accuracy, significantly outperforming traditional methods like MUSIC (78% accuracy) and ESPRIT algorithms [12, 17] (Fig. 7).

The Root Mean Square Error (RMSE) analysis shows consistent improvements: IoMT-Net achieves an average RMSE of 0.186 across all SNR levels compared to 0.523 for the MUSIC algorithm, demonstrating superior localization precision [12].

Table 1 presents a comprehensive comparison of computational complexity across different CNN architectures:

Table 1 Comparison of computational complexity and performance of CNN architectures

Architecture	Parameters	MFLOPs	Inference time (ms)	Accuracy at +5 dB SNR (%)
LRCANet	215 K	103	0.0042	96.35
IoMT-Net	142 K	142	0.417	97.63
ResNet	990 K	397	0.0055	90.85
SqueezeNet	555 K	332	0.0062	90.83
Multi-CNN	67.51 M	342	0.0074	88.23
Broadband-CNN	67.23 M	274	0.0053	90.70

The experimental results demonstrate that both proposed architectures achieve superior trade-offs between accuracy and complexity compared to existing methods. LRCANet achieves 96.35% accuracy with only 215 K parameters and 103 MFLOPs, while IoMT-Net reaches 97.63% accuracy with 142 K parameters, making both suitable for deployment on resource-constrained UAV platforms [12, 17].

4 Integrated Blockchain-AI Architecture

The convergence of blockchain technology with artificial intelligence creates a synergistic security framework that addresses the multifaceted challenges inherent in Internet of Drones (IoD) networks. This section presents the comprehensive, integrated architecture that seamlessly combines the decentralized trust mechanisms of blockchain with the real-time threat-detection capabilities of lightweight machine learning (ML) models, establishing a robust, scalable, and autonomous security ecosystem for UAV operations.

4.1 AI Integration Framework

The proposed integrated architecture extends beyond conventional CNN-based detection systems by incorporating language models, enabling sophisticated, context-aware threat analysis. The BLM-Chain framework represents a significant advancement in IoD security by addressing the limitations of traditional signature-based detection methods that rely on predefined patterns and behaviors. While conventional approaches achieve effectiveness against recognized attacks, they demonstrate reduced capability in detecting novel or sophisticated threats that exploit emerging vulnerabilities in UAV communication systems [15]. The architectural integration leverages the complementary strengths of multiple AI paradigms:

- Convolutional Neural Networks (CNNs): Perform deep feature extraction from RF signals and network traffic patterns
- Large Language Models (LLMs): Provide context-aware threat analysis and semantic understanding of complex attack vectors
- Blockchain Infrastructure: Ensures immutable audit trails, decentralized consensus, and automated policy enforcement through smart contracts.

4.2 Real-World Application Scenarios

4.2.1 Blockchain-Secured Autonomous Delivery Fleet Operations

The urban drone delivery ecosystem presents complex coordination challenges involving multiple stakeholders, including e-commerce platforms, logistics providers, recipients, regulatory authorities, and insurance companies [24, 25]. The proposed blockchain-AI framework addresses these challenges through end-to-end transparency and automated trust establishment, eliminating intermediaries while maintaining accountability.

When a customer initiates a delivery request through an e-commerce platform, a smart contract is instantiated on the blockchain that encodes the complete delivery terms including pickup location coordinates, destination geospatial data, delivery time window constraints, package authentication parameters, and payment escrow conditions. The e-commerce platform's fleet management system selects an available delivery UAV from its fleet based on proximity to the pickup location, current battery charge level, payload capacity constraints, and historical performance metrics. The selected UAV's identity is cryptographically verified against the blockchain registry through digital signature validation, confirming valid aircraft registration with aviation authorities, active insurance coverage meeting minimum liability requirements, and operator licensing credentials.

Upon package acquisition at the fulfillment center, Internet of Things sensors embedded in the secure delivery container record package weight, three-dimensional measurements, seal integrity status, and environmental conditions, including temperature and humidity, for sensitive items. This sensor data is written to the blockchain as an immutable record of the custody transfer, with cryptographic timestamps, establishing a verifiable chain of custody that protects against liability disputes. The multi-signature access control mechanism requires cryptographic signatures from both the warehouse operator and the UAV to authorize package release, preventing unauthorized access to delivery containers.

Throughout the flight, the UAV periodically broadcasts encrypted position updates via cellular or satellite links, which are validated against the authorized flight corridor defined in the smart contract. The AI-driven threat detection system continuously monitors the surrounding airspace for unauthorized UAVs that may pose collision risks or security threats, triggering automated collision avoidance protocols and logging incident reports to a blockchain accessible to regulatory authorities. The system employs real-time deconfliction algorithms powered by artificial intelligence that analyze airspace traffic density, weather forecast data, ground risk assessments, and vehicle performance parameters to adapt flight paths as conditions change autonomously.

When approaching the delivery destination, the smart contract executes geofence validation protocols to confirm the UAV remains within authorized operational airspace and has not deviated into restricted zones such as airports, military installations, or critical infrastructure protection areas. Recipient authentication occurs via zero-knowledge proof protocols, in which the recipient demonstrates possession of a cryptographic credential linked to the order without revealing personal identity information that could compromise privacy. Upon successful authentication and package handoff verification through container tamper sensors detecting seal breakage, the smart contract automatically releases escrowed payment funds to the delivery operator and records completion timestamp with delivery confirmation signatures on the immutable blockchain ledger.

This comprehensive audit trail provides complete transparency for all stakeholders, including the customer, who can track delivery progress in real time; the platform operator, monitoring fleet performance; insurance providers, assessing risk profiles; and regulatory authorities, conducting compliance audits, while protecting individual privacy through selective disclosure mechanisms and encryption. The blockchain architecture enables autonomous coordination across competing delivery fleets through standardized smart contract interfaces that facilitate interoperability. When a delivery operator experiences fleet capacity constraints during peak demand periods, overflow orders can be automatically brokered to partner operators through inter-contract transactions with service fees and liability terms cryptographically encoded in the agreement, creating a collaborative ecosystem that improves overall service reliability.

4.2.2 Emergency Response Operations

Natural disaster scenarios, including earthquakes, floods, hurricanes, and wildfires, demand rapid coordinated UAV deployment for search and rescue operations, combined with critical medical supply delivery to isolated victims in areas where traditional ground transportation infrastructure has been destroyed or rendered impassable. Traditional centralized command structures suffer from communication infrastructure failures during disasters, leading to coordination breakdowns and delayed response times that directly impact survival rates. The decentralized blockchain framework maintains operational continuity even when ground infrastructure, including cellular towers and internet backbone connections, is compromised through distributed consensus mechanisms and mesh network topologies.

When a disaster strikes and emergency operations are activated, response coordinators deploy reconnaissance UAVs equipped with thermal imaging sensors for detecting heat signatures of survivors, LiDAR systems for three-dimensional terrain mapping, and high-resolution cameras for structural damage assessment. These UAVs operate as blockchain nodes with onboard computing capabilities, maintaining distributed consensus on discovered victim locations with precise GPS coordinates, victim status assessments including mobility and consciousness indicators, and navigational hazards including collapsed structures and flooded areas. Each discovery generates a blockchain transaction containing encrypted coordinates to prevent adversarial exploitation, victim triage severity scores based on observed vital signs, and environmental conditions, including ambient temperature and structural stability.

The AI-driven threat detection system identifies potential secondary hazards, such as unstable building structures at risk of collapse, downed power lines carrying electrical hazards, chemical spill zones, and hostile environmental conditions, including fire perimeters and floodwater flow patterns. These hazards are logged as dynamic no-fly zones and immediately propagated across all participating UAVs via blockchain consensus, preventing inadvertent exposure to dangerous conditions. Medical supply delivery UAVs carrying essential items, including automated external defibrillators, trauma kits, emergency medications, blood products, and water purification supplies, access the victim location database via smart contract queries that automatically prioritize deliveries based on triage severity scores.

The blockchain-based coordination prevents multiple UAVs from redundantly servicing the same victim location while ensuring comprehensive coverage of all identified victim sites through deterministic task allocation algorithms implemented in smart contracts. When a medical UAV confirms successful delivery of supplies through automated package-drop verification sensors, it updates the victim status record on the blockchain with the delivery confirmation timestamp and the contents delivered, triggering real-time notifications to ground rescue teams regarding the specific medical conditions treated and the recommended intervention priorities for follow-up care. This information flow enables rescue personnel to optimize their response by prioritizing victims requiring immediate evacuation versus those stabilized with delivered supplies who can await secondary extraction.

The reputation-based trust scoring mechanism proves critical in disaster scenarios where UAVs from multiple agencies, including federal emergency management, military units, non-governmental organizations, and volunteer organizations, must coordinate without prior operational experience. Each participating UAV's credentials, including operator certifications, equipment capabilities, and historical performance metrics, are cryptographically verified against the blockchain registry with trust scores computed from past emergency response participation. High-reputation UAVs assigned to critical supply deliveries carrying time-sensitive medications, including insulin for diabetics or epinephrine for anaphylaxis, receive priority corridor access and expedited clearance through congested airspace where multiple aircraft compete for limited flight paths. The system automatically filters out unauthorized or unverified UAVs operated by media organizations or curious observers that may interfere with official operations, logging intrusion attempts as security incidents and reporting them to command authorities.

Blockchain immutability provides crucial accountability for post-incident analysis, enabling systematic evaluation of response effectiveness and identification of process improvements for future emergency responses. Complete flight logs, including three-dimensional trajectories, delivery confirmations with package contents and timestamps, victim discovery records, and response timelines, are permanently recorded in the distributed ledger with cryptographic integrity protection. Insurance claims processing for damaged equipment, liability disputes over delayed responses, and performance evaluations for participating organizations are supported by objective, verifiable data extracted from the blockchain, eliminating disputes based on conflicting accounts. The decentralized architecture ensures that mission-critical data survives even if individual blockchain nodes are physically damaged or electronically compromised during a disaster, thanks to data replication across geographically distributed nodes.

4.2.3 Cooperative Aerial Surveillance with Multi-agency Blockchain Trust Scoring

Border security operations, critical infrastructure protection including power plants and water treatment facilities, and large-scale public event monitoring require coordinated surveillance involving UAVs from multiple agencies with varying trust levels, authorization scopes, and operational mandates. The blockchain framework establishes a federated surveillance network where participating UAVs share situational awareness data while maintaining operational security through cryptographic access controls and multi-level security classifications.

Each surveillance UAV registers its operational parameters on the blockchain ledger, including sensor capabilities such as camera resolution and field of view, authorized monitoring zones defined by geographic boundaries and altitude limits, clearance level for accessing classified information, and inter-agency data sharing permissions encoded as smart contract access control lists. Multi-signature smart contracts define collaborative surveillance missions in which UAVs from different agencies must jointly validate anomalous observations before escalating alerts to command centers, thereby reducing false-positive rates that waste response resources. For instance, when a border patrol UAV detects suspicious activity, such as unauthorized border crossings, the observation is recorded as a blockchain transaction that requires confirmation from neighboring surveillance units operated by customs enforcement or the coast guard before triggering official interdiction response protocols.

This multi-validator approach reduces false-positive alert rates by requiring consensus among independent sensors with different observation angles and sensor modalities, while maintaining distributed decision authority to prevent single points of failure. The system implements game-theoretic coordination strategies in which UAVs negotiate surveillance responsibilities for overlapping zones via smart contract-mediated auctions, optimizing coverage while minimizing redundancy and conserving limited battery resources. Financial transactions recorded on the blockchain enable budget-tracking, allowing agencies to compensate partner organizations for providing surveillance coverage in jurisdictions outside their primary responsibility.

The AI-driven intruder detection system operates continuously across all surveillance UAVs, identifying unauthorized drones entering protected airspace using radar signatures, visual detection, and radio-frequency signature analysis. When an intrusion is detected, the three-dimensional localization data, including position coordinates, velocity vector, and altitude, is immediately broadcast across the blockchain network through consensus transactions, alerting all participating surveillance units to the emerging threat. The cooperative tracking protocol enables triangulation of the intruder's position from multiple simultaneous UAV observations at different vantage points, with consensus algorithms filtering measurement noise and sensor uncertainties to improve localization accuracy via Kalman and particle filtering.

Intrusion response follows predefined escalation protocols encoded in smart contracts that implement graduated intervention levels based on automated threat assessment scores computed from intruder behavior patterns. Low-threat incursions, such as recreational drones inadvertently straying into restricted airspace, trigger automated warning messages instructing the operator to exit the protected zone immediately. Medium-threat scenarios involving deliberately evasive behavior trigger coordinated pursuit protocols, in which multiple surveillance UAVs converge on the intruder to maintain continuous tracking while law enforcement is dispatched. High-threat situations, such as multiple coordinated intruders suggesting an organized attack, trigger immediate notification to national security authorities, with complete sensor data streams forwarded for real-time analysis.

Trust-scoring mechanisms govern data-sharing privileges among participating agencies based on organizational clearance levels and operational reliability. UAVs operated by trusted federal intelligence agencies with top-secret clearances receive full access to collaborative surveillance data, including raw sensor feeds and analytical products, while contractor-operated UAVs face restricted access limited to their specific operational zones and sanitized data products with sensitive information redacted. The blockchain maintains immutable audit logs of all data access events, with cryptographic signatures that identify the accessing entity, timestamp, specific data retrieved, and authorized purpose, ensuring complete accountability and preventing unauthorized intelligence gathering that violates inter-agency agreements.

When surveillance data reveals security-sensitive information such as the locations of undercover operatives or technical capabilities of detection systems, ring signature protocols enable authorized personnel to annotate observations without revealing their specific organizational affiliation, protecting operational security while maintaining information quality through verifiable credentials. The blockchain architecture facilitates seamless handoff of surveillance responsibility as mobile targets transit across jurisdictional boundaries between different agency areas of responsibility. Smart contracts automatically transfer tracking authority to the appropriate agency based on geographic zones encoded in the blockchain, with complete surveillance histories, including sensor data and analytical assessments, transferred via encrypted blockchain transactions that maintain a chain of custody. This capability is essential for sustained monitoring of mobile targets, such as suspected smugglers or terrorists, moving across extended geographic regions involving multiple oversight authorities with complex jurisdictional frameworks.

5 Smart Contract Automation and Response Mechanisms

5.1 BLM-Chain Framework Architecture

The BLM-Chain framework introduces a novel three-tier integration model that combines blockchain's immutable ledger capabilities with LLMs' advanced threat detection mechanisms [15]. The architecture begins with the comprehensive acquisition of network traffic from IoT devices and UAV sensors, where transmitted data, including vital communication between UAVs, Ground Control Stations (GCS), and other entities, is promptly recorded on the blockchain with timestamps, ensuring data integrity and maintaining verifiability (Fig. 8).

Upon threat detection, the integrated system utilizes blockchain-based smart contracts to automate response protocols. These contracts are pre-programmed with specific countermeasures, including:

- Isolation of compromised UAVs from network operations
- Dynamic rerouting of communication pathways to bypass affected nodes

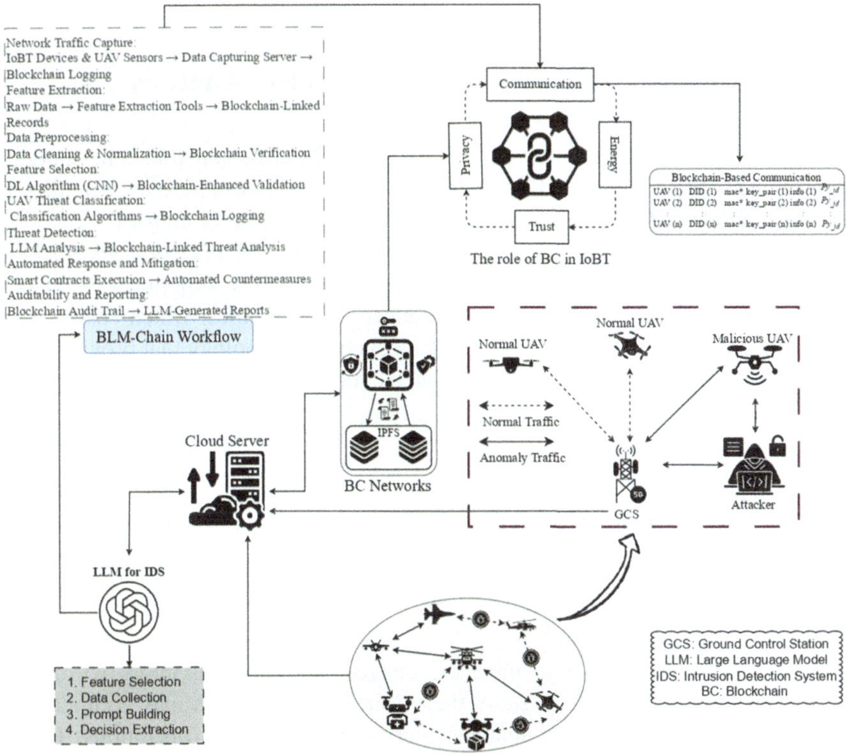

Fig. 8 Architecture of the proposed BLM-chain threat resistance framework

Table 2 Performance comparison of integrated blockchain-AI architectures for UAV threat detection

Model	Accuracy (%)	Precision (%)	Recall (%)	F1-score (%)	Architecture type
ANN-based	93.43	93.09	89.92	92.35	Centralized
ML-based	95.46	93.68	90.68	90.06	Centralized
FL-based	93.37	90.86	94.56	92.29	Distributed
BLM-chain	96.12	93.36	95.15	93.85	Decentralized

- Activation of defensive protocols against identified threat vectors
- Real-time alerting of nearby UAVs and ground control stations.

The automation provided by smart contracts ensures that responses execute within milliseconds without manual intervention, which is crucial for the fast-paced dynamics of the battlefield. Every action taken by the system is securely logged on the blockchain, providing comprehensive, immutable audit trails for post-incident analysis and accountability verification.

6 Performance Evaluation and Comparative Analysis

An experimental evaluation using the 5G-NIDD dataset demonstrates the superior performance of the integrated BLM-Chain architecture compared to existing threat-resistance models. Table 2 presents comprehensive performance metrics:

The BLM-Chain model achieves 96.12% accuracy, 93.36% precision, and 95.15% recall, significantly outperforming traditional approaches. This superior performance stems from the LLM's advanced feature selection capabilities, which ensure only relevant data is processed, combined with blockchain technology that provides data authenticity and integrity verification, crucial for maintaining high precision and recall rates.

7 The Integration of Privacy-Preserving Blockchain in IoD

7.1 Privacy-Preserving Blockchain Mechanisms for IoD

While blockchain technology enhances transparency and auditability within the Internet of Drones (IoD) ecosystem, these same characteristics may conflict with the privacy requirements of unmanned aerial vehicle (UAV) operations. Public ledgers inherently expose transaction metadata such as UAV identifiers, timestamps, and operational zones, which can inadvertently reveal sensitive information about flight paths, operator identities, and mission objectives [26, 27]. Addressing these vulnerabilities requires integrating advanced privacy-preserving mechanisms into blockchain-based IoD frameworks. This subsection elaborates on three principal techniques: zero-knowledge proofs (ZKPs), differential privacy (DP), and ring signatures (RS), and their applicability in IoD environments.

7.1.1 Zero-Knowledge Proofs (ZKPs)

Zero-knowledge proofs allow a prover to demonstrate the validity of a statement without disclosing the underlying data [28]. In blockchain systems, non-interactive variants such as zk-SNARKs and zk-STARKs have been employed to verify transactions or identity credentials while concealing raw information. Within IoD networks, ZKPs can validate UAV compliance with geofencing or altitude constraints (e.g., "the UAV remained within the assigned operational zone") without revealing its precise coordinates. This ensures auditability through blockchain verification while protecting the UAV's operational confidentiality. Moreover, ZKPs can secure smart contracts, enabling access control or policy enforcement without disclosing private mission parameters [27].

However, ZKP implementation poses computational challenges, particularly in UAVs with limited onboard resources. Efficient deployment requires lightweight schemes or offloading proof generation to edge servers, leaving on-chain verification for decentralized validation. Emerging ZKP frameworks tailored for resource-constrained environments offer promising directions for future IoD systems.

7.1.2 Differential Privacy (DP)

Differential privacy introduces controlled noise to data outputs, ensuring that the inclusion or exclusion of any single record has minimal effect on analytical results [29]. This technique is valuable when aggregated UAV telemetry or performance metrics must be shared publicly or stored on the blockchain. For example, DP can protect operator anonymity in fleet-level analyses of flight density or communication patterns. In an IoD blockchain, differential privacy mechanisms can be applied to off-chain data aggregations, thereby balancing the trade-off between privacy preservation and data utility.

Recent studies have explored the fusion of verifiable DP and ZKP methods, allowing a blockchain to confirm that data perturbations satisfy differential privacy guarantees without exposing raw values [29]. Such a combination is particularly relevant for IoD analytics, where mission-sensitive data must remain confidential but verifiably compliant with privacy policies.

7.1.3 Ring Signatures (RS)

Ring signatures enable a member of a predefined group to sign a message such that the verifier can confirm that one member of the group signed it, without revealing which one [30]. This cryptographic construct offers anonymity and unlinkability—crucial for UAV operator privacy. In a blockchain-enabled IoD network, RS can be used to anonymize UAV or operator identities in transaction logs, e.g., a maintenance report signed by "one of the certified fleet operators" rather than by a specific UAV ID.

This technique preserves auditability while mitigating identity exposure risks. Recent works on elliptic-curve-based ring signatures improve efficiency and reduce verification latency, making them suitable for IoD environments with limited resources [31]. Nonetheless, anonymity guarantees depend on the size and composition of the ring; small anonymity sets may enable deanonymization attacks, while large sets increase verification overhead. Hybrid schemes combining RS with DP can balance these factors, offering formal privacy bounds (e.g., ε-indistinguishability) for blockchain records [30].

7.1.4 Integration Within the IoD Architecture

Integrating these privacy-preserving mechanisms strengthens the resilience of blockchain-based IoD systems:

- **ZKPs** validate UAV compliance with mission policies and geofencing constraints without revealing sensitive telemetry data.
- **Ring Signatures** anonymize UAV/operator identities in blockchain transactions while maintaining verifiable accountability.
- **Differential Privacy** safeguards statistical UAV data sharing and analytics by introducing calibrated noise to aggregated datasets.

To minimize computational overhead, ZKP generation can be delegated to Zone Edge Servers (ZES), while verification remains on-chain. Similarly, DP mechanisms may be applied at the data preprocessing layer before blockchain storage. Table 3 summarizes the comparative advantages and limitations of these approaches.

By adopting these complementary mechanisms, IoD frameworks can achieve a robust balance between transparency, auditability, and privacy. Such integration transforms the blockchain from a merely secure ledger into a privacy-aware infrastructure capable of supporting trustworthy, regulation-compliant, and zero-trust UAV ecosystems.

Table 3 Comparative analysis of privacy-preserving techniques in blockchain-enabled IoD

Technique	Primary function	Advantages	Limitations
Zero-knowledge proofs	Policy and identity validation without data exposure	Strong confidentiality; verifiable compliance	High proof-generation cost; circuit design complexity
Differential privacy	Statistical data privacy via noise addition	Protects aggregate UAV metrics; tunable privacy budget	Loss of data accuracy; limited protection for individual transactions
Ring signatures	Identity and signer anonymity within a group	Lightweight authentication; unlinkability	Requires large anonymity sets for strong privacy; group management complexity

8 Blockchain-Specific Risks and Security Analysis and Discussion

8.1 Consensus Attacks and Byzantine Fault Tolerance in IoD Networks

The proposed Delegated Proof-of-Authority (DPoA) consensus mechanism, while energy-efficient and optimized for IoD applications with average block times of approximately 1–2 s, introduces specific attack vectors that must be carefully addressed. Byzantine fault tolerance is the foundational requirement for blockchain networks to reach consensus even when some nodes exhibit malicious behavior, either through intentional attacks or unpredictable faults. The primary vulnerability in authority-based consensus mechanisms stems from the concentration of validation authority among a designated set of validator nodes. If adversaries compromise a sufficient number of validators through coordinated attacks, credential theft, or insider threats, they could potentially manipulate transaction ordering, exclude legitimate UAV authentication requests, inject fraudulent flight authorizations, or censor critical safety alerts.

Byzantine fault tolerance theory establishes that consensus mechanisms can tolerate malicious behavior from at most one-third of validators while maintaining safety and liveness properties. For a UAV network with 30 validators, this means the system remains secure as long as fewer than 10 validators exhibit Byzantine behavior. The DPoA implementation in the proposed framework incorporates multiple defensive layers to fortify this theoretical tolerance. First, validator selection employs strict geographic and organizational diversity requirements, ensuring that validators are distributed across multiple physical locations, jurisdictional boundaries, and independent operating organizations. This spatial distribution prevents single-point compromises or coordinated cyber-attacks on a specific geographic region from cascading across the entire validator set.

Second, dynamic validator rotation protocols periodically shuffle validation responsibilities among a larger pool of authorized nodes, typically cycling validator sets every 1,000 blocks or approximately 15 min of blockchain operation. This rotation mechanism limits the temporal window during which a compromised validator can inflict damage and distributes risk across the broader validator pool. An attacker who compromises a single validator node would gain validation authority for only the rotation period before being replaced, limiting accumulated damage. Third, real-time monitoring mechanisms continuously track validator behavior patterns against baseline profiles, detecting anomalies such as excessive transaction rejections deviating from statistical norms, suspicious proposal patterns suggesting transaction manipulation, unusual communication patterns indicating validator collusion, or timing irregularities suggesting Byzantine behavior.

The hierarchical architecture with Zone Edge Servers provides additional resilience through local consensus domains that operate independently of central consortium chain validation. Intra-zone UAV coordination transactions are validated entirely within local zone validators, ensuring operational continuity for critical safety functions even if central chain validators are subjected to Byzantine attacks. Cross-zone operations may experience temporary authorization delays during consensus attacks on the central chain, but safety-critical autonomous functions including collision avoidance, emergency descent, and obstacle avoidance continue operating through local decision-making without blockchain dependency.

8.1.1 Attack Vectors and Mitigation in Consortium Blockchain Deployments

While traditional 51% attacks targeting public blockchains focus on computational resource accumulation through mining hardware control, consortium chain implementations face fundamentally different attack dynamics centered on validator collusion, credential compromise, and organizational infiltration. In the proposed IoD framework deployed as a permissioned consortium blockchain, a successful 51% attack would require adversaries to control more than half of the authorized validator nodes, achieving consensus capability sufficient for malicious transaction approval or ledger manipulation.

The economic and operational barriers to executing such attacks are substantial in practice. Validator nodes in the proposed framework are operated by trusted entities including military aviation commands, government aviation regulatory agencies, and nationally certified commercial UTM operators with significant reputational stakes in system integrity and operational continuity. The multi-stakeholder validator composition ensures that no single organization or nation-state controls sufficient validators to achieve majority consensus for malicious purposes unilaterally. Additionally, validator authorization requires cryptographic credentials issued by multiple independent certification authorities with separate approval hierarchies, preventing single-authority compromise from granting attackers validator node access.

However, sophisticated adversaries with nation-state capabilities could potentially compromise multiple validators through coordinated cyber-attacks exploiting software vulnerabilities, credential theft from key management systems, social engineering targeting validator operators, or supply chain exploitation introducing compromised components into validator infrastructure. To address this residual risk, the system implements continuous integrity monitoring that analyzes blockchain state consistency across all geographically dispersed validator nodes. Ledger divergences that indicate potential chain reorganization attacks trigger immediate automated security alerts and initiate credential revocation procedures for suspected compromised validators. The system automatically quarantines divergent validator nodes from the active consensus set, requiring successful security audit and credential re-issuance before reactivation.

Critical operational parameters subject to potential attack including emergency airspace closures affecting large geographic regions, modification of geofence boundaries for restricted zones, or authorization level adjustments for specific operators are subject to dual-consensus requirements. These transactions require independent validation from both local zone validator quorum and a geographically diverse set of central consortium validators, preventing single-domain compromise from affecting system-critical policies. The computational cost of generating 51% attacking power in a 30-validator system requires compromising at least 16 validator nodes, each operating on hardened infrastructure with cryptographic key storage, access logging, and multi-party authorization controls. This represents substantially higher costs and complexity than traditional mining attacks.

The system incorporates cryptographic checkpointing mechanisms in which the blockchain state is periodically signed by trusted third-party auditors with no direct validator authority but substantial institutional credibility and an incentive for accuracy. State checkpoints are generated every 10,000 blocks, approximately every 2.8 h, with cryptographic commitments produced by independent auditors and published via multiple secure channels. These checkpoint signatures provide verifiable reference points that limit the temporal scope of potential chain reorganization attacks. Even if adversaries achieve temporary majority control over active validators, they cannot retroactively alter the transaction history beyond the most recent checkpoint without detection by comparing it against the independently generated cryptographic commitment. This approach combines the permissioned efficiency of consortium blockchains with the historical immutability guarantees of blockchain technology.

8.1.2 Latency Implications in Real-Time Drone Swarms and Mitigation Strategies

Real-time UAV coordination demands sub-second decision latency for critical safety functions, including reactive collision avoidance, emergency maneuvers in response to sensor detections, and dynamic obstacle response when navigating around detected hazards. Traditional blockchain consensus protocols introduce inherent latency penalties due to block propagation delays as blocks are transmitted across distributed validator nodes, the validation processing time required for each validator to execute smart contracts and verify transaction signatures, and finality confirmation requirements that require multiple sequential blocks before transaction reversions become cryptographically impractical.

In the proposed DPoA implementation optimized for UAV operations, block mining times average approximately 1–2 s with standard deviation less than 200 ms, providing deterministic latency characteristics suitable for UAV traffic planning and airspace coordination decisions. However, this latency approaches but does not fully satisfy ultra-low-latency requirements for certain swarm coordination scenarios requiring sub-100 ms responsiveness. For example, autonomous collision avoidance algorithms need detection-to-maneuver latency below 50 ms to maintain safe separation at typical UAV speeds exceeding 15 m/s.

The framework addresses these latency constraints through a hybrid on-chain/off-chain architecture that stratifies coordination functions by latency sensitivity and safety criticality. Time-critical operations, including immediate collision avoidance through reactive trajectory adjustments, emergency manoeuvres in response to ground sensor alerts, and reactive altitude or heading changes, operate via direct peer-to-peer communication channels between UAVs using cryptographically authenticated messages. These off-chain coordination protocols leverage shared secrets established during initial blockchain-based UAV registration and authorization, enabling secure direct communication without requiring per-message blockchain validation. Experimental evaluations indicate that peer-to-peer authenticated communication achieves latencies of 10–50 ms, depending on the communication medium and link quality, substantially lower than blockchain confirmation times.

Less time-sensitive operations, including strategic flight plan registration and approval, geofence modification transactions, long-duration corridor reservations, and immutable audit logging for regulatory compliance, occur through standard blockchain transactions with full consensus validation. These operations tolerate blockchain latency of 1–5 s, since flight plan changes are typically planned with sufficient advance notice rather than requiring an immediate reaction.

For scenarios requiring both real-time coordination and blockchain verifiability for accountability, the system employs layer-two state channel technology, in which pairs or small groups of cooperating UAVs establish temporary off-chain coordination channels for the duration of collaborative operations. High-frequency position updates and coordination messages are exchanged through state channels with microsecond-scale latencies, while cryptographically committed summary state snapshots are written to the blockchain at lower frequencies. In the event of channel disputes regarding transaction sequence or state validity, the complete interaction history can be reconstructed from cryptographically committed channel state and adjudicated through on-chain smart contract arbitration. This hybrid approach maintains blockchain verifiability while achieving the latency performance necessary for real-time UAV operations.

Drone swarm coordination scenarios involving dozens of cooperating UAVs employ hierarchical consensus structures in which designated swarm leaders aggregate coordination messages from swarm members and submit batch transactions to the blockchain, rather than requiring individual blockchain transactions per UAV. This batching approach reduces blockchain transaction load by factors of 10 to 50, depending on swarm size, while maintaining individual UAV accountability through cryptographic signature chains that cryptographically link swarm-level transactions to constituent member actions. Each swarm member digitally signs its coordination data, which the leader bundles into a single blockchain transaction with all member signatures, enabling post-facto verification of each member's contribution through signature validation.

Performance analysis of the proposed hybrid architecture indicates that time-critical coordination operations achieve average latencies of 15 ms for off-chain peer-to-peer communication with cryptographic authentication, 12.4 ms average with standard deviation of 3.2 ms for state channel transactions as measured in recent blockchain-integrated drone swarm studies, and approximately 1.2–2.0 s for persistent state updates requiring full blockchain consensus finality. This performance profile satisfies requirements for typical UAV coordination scenarios in urban airspace with moderate to high traffic density while acknowledging that certain specialized applications including autonomous racing or high-speed evasive maneuvers may require purely off-chain coordination with post-operation blockchain reconciliation for audit and accountability purposes.

The design maintains safety-critical autonomy by ensuring that time-sensitive decision-making does not depend on blockchain confirmation, while leveraging the blockchain's immutability and auditability for accountability, regulatory compliance, and post-incident analysis. This layered approach balances conflicting requirements of real-time responsiveness and permanent auditability that would otherwise be irreconcilable if blockchain validation were required for every operational decision.

8.2 Discussion

- The convergence of blockchain and AI within Internet of Drones (IoD) networks has demonstrated pronounced advantages in enhancing the resilience of UAV operations against evolving cyber-physical threats. Experimental results confirm that integrating lightweight CNN models for real-time unauthorized UAV localization with permissioned blockchain ledgers and smart contracts yields high detection accuracy (>97%), sub-second decision latency, and scalable transaction throughput (>280 TPS). Moreover, the hierarchical blockchain architecture and delegated Proof-of-Authority consensus minimize energy consumption and single-point-of-failure (SPoF) vulnerabilities while maintaining Byzantine fault tolerance.
- Despite these advances, several challenges persist. First, the reliance on fixed non-uniform antenna arrays for direction-of-arrival estimation limits adaptability to varying flight configurations and dynamic environments. Future work should investigate adaptive antenna selection and beamforming strategies that dynamically optimize array geometry to achieve improved spatial resolution. Second, while lightweight CNN architectures achieve favorable accuracy-complexity

trade-offs, their fixed model parameters may degrade performance when confronting unforeseen signal distortions or adversarial manipulations. Integrating online learning techniques, such as federated or continual learning, would enable the AI modules to update their parameters in situ, thereby enhancing robustness against zero-day threats without requiring central retraining.
- Third, blockchain storage and transaction costs remain a non-negligible overhead for continuous logging of high-frequency detection events. Hybrid architectures that offload bulk telemetry to decentralized file systems (e.g., IPFS) or employ state-channel mechanisms could reduce on-chain storage demands while preserving auditability. Privacy-preserving enhancements such as zero-knowledge proofs and secure multi-party computation are also essential for protecting sensitive UAV telemetry and mission data, particularly in civilian or cooperative multi-stakeholder scenarios.
- Fourth, interoperability across heterogeneous UAV platforms and regulatory domains poses integration challenges. Standardized data formats, cross-chain protocols, and semantic ontologies are required to ensure seamless collaboration between civilian, commercial, and defense UAV fleets. Regulatory compliance mechanisms embedded directly within smart contracts could automate authorization workflows in accordance with evolving airspace policies, improving transparency and reducing administrative burdens.
- Finally, the emergence of more powerful AI paradigms, such as graph neural networks for anomaly correlations and transformer-based models for contextual threat interpretation, offers new opportunities to augment detection capabilities. Integrating these advanced AI models within the blockchain framework will require careful optimization to balance computational demands against latency requirements. Edge-cloud continuum architectures, leveraging split-computing and hardware accelerators, can accommodate these complex models while maintaining real-time responsiveness.

In summary, the integrated blockchain-AI framework provides a robust foundation for securing IoD networks; however, continued innovation is necessary to address concerns regarding adaptability, scalability, privacy, and interoperability. Advancing toward autonomous, self-evolving security architectures will involve synergistic enhancements in adaptive sensing, online learning, decentralized storage, policy automation, and the integration of next-generation AI, paving the way for resilient and trustworthy airspace ecosystems.

9 Conclusion and Future Directions

The convergence of permissioned blockchain and edge-optimized artificial intelligence (AI) has yielded a resilient security fabric capable of safeguarding dynamic Internet of Drones (IoD) deployments. By distributing trust across hierarchical ledgers with delegated Proof-of-Authority (PoA) consensus, ensuring sub-second transaction finality, and embedding smart contracts for autonomous geofence enforcement, the architecture eliminates single points of failure (SPoF) and prevents unauthorized incursions. Concurrently, lightweight convolutional neural networks deployed on UAV and edge server platforms deliver real-time angle-of-arrival estimation and intruder localization with over 96% accuracy, even in low-signal environments. Together, these technologies form a zero-trust framework in which every flight operation is cryptographically authenticated, continuously monitored, and immutably recorded.

Despite these achievements, the field must address emerging challenges to realize fully autonomous, self-evolving IoD security ecosystems. Adaptive antenna configurations and beamforming techniques can enhance detection robustness in complex terrain or urban canyons. At the same time, federated and continual learning paradigms will allow AI models to evolve in response to novel adversarial tactics without central retraining. Scalable on-chain/off-chain data management, leveraging state channels or decentralized storage networks, will be essential to balance auditability with storage efficiency, especially under high event frequencies. Integrating privacy-preserving cryptographic primitives, such as zero-knowledge proofs, will protect sensitive telemetry and mission data without impairing verifiability.

Looking forward, the next generation of IoD security architectures will integrate graph neural networks for multi-agent anomaly correlation and transformer-based models for contextual threat interpretation with the existing blockchain backbone. Standardizing cross-chain interoperability protocols and regulator-embedded compliance smart contracts will facilitate seamless collaboration across civilian, commercial, and defense UAV networks. By pursuing these research directions, the community can advance toward autonomous, self-adapting security frameworks that not only guard the skies today but also anticipate and neutralize the cyber-physical threats of tomorrow.

References

1. Boccadoro, P., Striccoli, D., Grieco, L.A.: An extensive survey on the internet of drones. Ad Hoc Netw. **122**, 102600 (2021)
2. Yahuza, M., Idris, M.Y.I., Ahmedy, I.B., Wahab, A.W.A., Nandy, T., Noor, N.M., Bala, A.: Internet of drones security and privacy issues: taxonomy and open challenges. IEEE Access **9**, 57243–57270 (2021)
3. Gerencheal, T.A., Lee, J.H.: Development of a collision avoidance system using multiple distance sensors for indoor inspection drone. Appl. Sci. **15**(16), 8791 (2025)
4. Lakshman, S.A., Ebenezer, D.: Integration of internet of things and drones and its future applications. Mater. Today: Proc. **47**, 944–949 (2021)
5. Abualigah, L., Diabat, A., Sumari, P., Gandomi, A.H.: Applications, deployments, and integration of internet of drones (IoD): a review. IEEE Sens. J. **21**(22), 25532–25546 (2021)
6. Munasinghe, I., Perera, A., Deo, R.C.: A comprehensive review of UAV-UGV collaboration: advancements and challenges. J. Sens. Actuator Netw. **13**(6), 81 (2024)
7. Sahraoui, S., Bachir, A.: Lightweight consensus mechanisms in the internet of blockchained things: thorough analysis and research directions. Digit. Commun. Netw. (2025)

8. Tsao, K.-Y., Girdler, T., Vassilakis, V.G.: A survey of cyber security threats and solutions for UAV communications and flying ad-hoc networks. Ad Hoc Netw. **133**, 102894 (2022)
9. Ajakwe, S.O., Kim, D.-S.: Facets of security and safety problems and paradigms for smart aerial mobility and intelligent logistics. IET Intel. Transport Syst. **18**, 2827–2855 (2024)
10. Ajakwe, S.O., Okoloegbo, C.A., Lee, J.M., Kim, D.S.: Machine learning models for drone security: cognitive versus cyber intelligence for safety operations. In: Machine Learning for Drone-Enabled IoT Networks: Opportunities, Developments, and Trends, pp. 121–139. Springer, Berlin (2025)
11. Ihekoronye, V.U., Ajakwe, S.O., Lee, J.M., Kim, D.-S.: Droneguard: an explainable and efficient machine learning framework for intrusion detection in drone networks. IEEE Internet Things J. (2024)
12. Akter, R., Golam, M., Doan, V.-S., Lee, J.-M., Kim, D.-S.: IOMT-net: blockchain-integrated unauthorized UAV localization using lightweight convolution neural network for internet of military things. IEEE Internet Things J. **10**(8), 6634–6651 (2022)
13. Akram, M.A., Ahmad, H., Mian, A.N., Jurcut, A.D., Kumari, S.: Blockchain-based privacy-preserving authentication protocol for UAV networks. Comput. Netw. **224**, 109638 (2023)
14. Alaba, F.A., Sulaimon, H.A., Marisa, M.I., Najeem, O.: Smart contracts security application and challenges: a review. Cloud Comput. Data Sci. 15–41 (2024)
15. Golam, M., Alam, M.M., Kim, D.-S., Lee, J.-M.: BLM-chain: Ai-driven blockchain for UAV threat resistance in IOBT. In: 2024 15th International Conference on Information and Communication Technology Convergence (ICTC), pp. 1609–1613. IEEE (2024)
16. Golam, M., Akter, R., Tuli, E.A., Kim, D.-S., Lee, J.-M.: Lightweight blockchain assisted unauthorized UAV access prevention in the internet of military things. In: 2022 13th International Conference on Information and Communication Technology Convergence (ICTC), pp. 890–894. IEEE (2022)
17. Golam, M., Akter, R., Naufal, R., Doan, V.-S., Lee, J.-M., Kim, D.-S.: Blockchain inspired intruder UAV localization using lightweight CNN for internet of battlefield things. In: MILCOM 2022-2022 IEEE Military Communications Conference (MILCOM), pp. 342–349. IEEE (2022)
18. Allouch, A., Cheikhrouhou, O., Koubâa, A., Toumi, K., Khalgui, M., Nguyen Gia, T.: UTM-chain: blockchain-based secure unmanned traffic management for internet of drones. Sensors **21**(9), 3049 (2021)
19. Akram, J., Anaissi, A.: DDRM: distributed drone reputation management for trust and reliability in crowdsourced drone services. In: 2024 IEEE International Conference on Web Services (ICWS), pp. 921–931. IEEE (2024)
20. Liu, Q., Dai, W., Ma, L., Tessone, C.J.: Towards transparent and privacy-preserving urban airspace management: a blockchain-based scheme under the airspace-resource-centric concept. In: 2025 Integrated Communications, Navigation and Surveillance Conference (ICNS), pp. 1–8. IEEE (2025)
21. Hossain, M.I., Tahtali, M., Turhan, U., Biswas, K.: Blockchain integration in UAV networks: performance metrics and analysis. Sensors **24**(23), 7813 (2024)
22. Arshad, U., Halim, Z.: Secure and optimized drone swarm operations with decentralized adaptive differential evolution. Comput. Electr. Eng. **126**, 110487 (2025)
23. Ajakwe, S.O., Olabisi, K.L., Kim, D.-S.: Multihop intruder node detection scheme (MINDS) for secured drones' FANET communication. IET Intel. Transport Syst. **19**(1), 70080 (2025)
24. Ajakwe, S.O., Saviour, I.I., Kim, J.-H., Kim, D.-S., Lee, J.M.: Banda: a novel blockchain-assisted network for drone authentication. In: 2023 Fourteenth International Conference on Ubiquitous and Future Networks (ICUFN), pp. 120–125. IEEE (2023)
25. Ajakwe, S.O., Igboanusi, I.S., Lee, J.-M., Kim, D.-S.: i BANDA: a blockchain-assisted defense system for authentication in drone-based logistics. Drones **9**(8), 590 (2025)
26. Garcia, R.D., Ramachandran, G., Dunnett, K.: A survey of blockchain-based privacy applications: an analysis of consent management and self-sovereign identity approaches (2024). arXiv:2411.16404

27. Lavin, R., Liu, X., Mohanty, H., Krishnamachari, B.: A survey on the applications of zero-knowledge proofs (2024). arXiv:2408.00243
28. Gupta, S.: Zero-knowledge proofs for privacy-preserving systems: a survey across blockchain, identity, and beyond. Eng. Technol. J. (2025)
29. A verifiable scheme for differential privacy based on zero-knowledge. Springer Open Computer Science (2025)
30. Wang, L., Chen, Y., Xu, P.: Secure ring signature scheme for privacy-preserving blockchain systems. Entropy **25**(9), 1334 (2023)
31. Saad, M., Ouyang, K., Singh, R.: A framework for enhancing privacy and anonymity in IoT/edge systems using ring signatures and zero-knowledge proofs. Elsevier Comput. Netw. (2024)

Security Threats, Cyber Risks, Smart Contracts, and Blockchain Governance for AIV

Threats and Security Challenges in Autonomous Intelligent Vehicles

Douglas Allswell Kelechi, Ikechukwu Ignatius Ayogu, Lucky Uchenna Chibuike, Chidi Ukamaka Betrand, and Ajere Ikenna Uzoma

Abstract The recent advances in autonomous intelligent vehicles (AIVs) research has increased the general awareness of its potential to reshape the outlook for the future of transportation. However, to be successful, AIVs must be able to navigate the complex and highly unpredictable driving environment at human-level accuracy. To satisfy this requirement, AIVs combine techniques of artificial intelligence, sensor fusion, and image processing, among others, for real-time decision-making in the course of driving. The integration of these technologies unavoidably exposes AIVs to traditional security vulnerabilities, primarily arising from networks and data communications protocols that are employed in AIVs. This paper conducts a systematic study of the traditional and emerging threat landscape for AIVs—cyber-physical threats, adversarial attacks, GPS spoofing, sensor jamming, and other plausible sources of breach to the AIV control systems. This paper also reviews software-associated risks, supply chain weaknesses, and insider threats, and analyzes the implications of these threats on safety, privacy, trust, and fidelity, and the wider adoption of AIVs. Finally, this paper evaluated the current best practices for mitigating the identified security

D. A. Kelechi (✉) · I. I. Ayogu · C. U. Betrand · A. I. Uzoma
Department of Computer Science, Federal University of Technology, Owerri, Imo State, Nigeria
e-mail: kelechi.douglas@futo.edu.ng

I. I. Ayogu
e-mail: ignatius.ayogu@futo.edu.ng

C. U. Betrand
e-mail: chidi.betrand@futo.edu.ng

A. I. Uzoma
e-mail: Ikenna.Ajere@futo.edu.ng

L. U. Chibuike · A. I. Uzoma
University of Agriculture and Environmental Sciences, Umuagwo, Imo State, Nigeria
e-mail: uchenna.lucky@uaes.edu.ng

A. I. Uzoma
Department of Information Technology, Federal University of Technology, Owerri, Imo State, Nigeria

S. O. Ajakwe and D.-S. Kim (eds.), *Blockchain for Security of Autonomous Intelligent Vehicles*, Studies in Big Data 194,
https://doi.org/10.1007/978-3-032-20851-4_6

issues for AIVs and suggested an adaptive security architecture for the evolving AIV technology landscape.

1 Introduction

Autonomous intelligent vehicles (AIVs) have revolutionized the automotive industry and possess the capacity to revolutionize conveyance by improving efficiency, safety as well as convenience. A network of sensors, software, and connectivity is necessary for self-driving cars, or AIVs, to navigate and function without human assistance. The safe and dependable operation of autonomous cars depends on addressing a number of cybersecurity issues, notwithstanding the many potential advantages. The interconnection of AIV frameworks creates a broad attack surface for potential cyber threats, as these systems rely on continuous data exchange among vehicles, infrastructure and external parties.

According to [1, 2], these attacks may target cloud-based services, communication networks, onboard systems, and other elements of the AIV ecosystem. Modern technologies like blockchain, AI, and machine learning have enhanced AIV performance, but they have also made it more vulnerable to potential hostile actor attacks. Because it relies on wireless communication between vehicles or infrastructure, the modern intelligent vehicle network is vulnerable to numerous threats. Additionally, its increasing connection has resulted in heightened security risks.

The complicated security environment of AIVs is made up of three main categories of possible threats: operational vulnerabilities, cyber hazards, and data privacy issues. Cybersecurity threats are especially relevant since AIVs are vulnerable to data breaches, hacking, and other types of cyberattacks because of their interconnection. An attacker may, for example, alter sensor data, gain unauthorized access to a car's systems, or interfere with the vehicle's ability to connect to other networks. By compromising the vehicle's performance and safety, such violations could pose a major risk to other drivers and passengers.

A thorough and proactive strategy is needed to solve the security issues in AIVs, one that incorporates strong cybersecurity safeguards, data protection procedures, and exacting testing and validation techniques. To find and address these weaknesses, secure communication protocols must be put in place, sensitive data must be encrypted, and frequent security audits and assessments must be carried out. Setting and enforcing strict security standards and best practices also requires cooperation from industry players, such as manufacturers, regulators, and cybersecurity specialists. Prioritizing security and taking a comprehensive approach to the dangers and difficulties that AIVs face will enable us to create more dependable and secure autonomous systems that live up to their potential of improving transportation efficiency, convenience, and safety.

To guarantee the security and dependability of these cutting-edge systems, the many security threats and difficulties that AIVs face, as well as possible hazards and mitigation techniques, will be covered.

What are the primary security threats to intelligent, self-driving cars? This book chapter attempts to address these issues. What impact do different security issues have on the security, safety, and dependability of autonomous intelligent vehicles? How can autonomous intelligent cars' present security and defenses be improved? How can the safety of intelligent, self-driving cars be guaranteed?

The study's goals on security concerns in autonomous intelligent vehicles are as follows:

1. To identify potential security threats
2. To look at vulnerabilities in autonomous car systems
3. To provide recommendations for improving security:

2 Cybersecurity Threats

Autonomous intelligent vehicles (AIVs) have many cybersecurity issues that need to be fixed to guarantee their reliable and safe functioning, notwithstanding its potential benefits. Cloud-based services, communication networks, and onboard systems are just a few of the AIV ecosystem's components that these threats may target [1, 3]. Passenger safety is compromised by a number of network security risks that might affect connected vehicles.

Several factors make AIV cybersecurity crucial. First of all, accidents, fatalities, and significant property loss could ensue from any interference with an AIV's control systems. Second, sensitive data theft or alteration raises serious privacy issues, including user preferences and location data. Third, AV intrusions could erode public confidence and prevent the technology from being widely used [4]. Strong cybersecurity measures in AIVs require a comprehensive approach that includes developing secure hardware and software architectures, using state-of-the-art encryption techniques, and continuously detecting and responding to threats. The development of industry standards and best practices requires cooperation between manufacturers, regulators, and cybersecurity specialists [5].

We examined cybersecurity risks and solutions in the developing field of autonomous intelligent vehicles in-depth and perceptively in this book chapter.

(i) Hacking remotely

Because they depend heavily on wireless connection for essential tasks, including data interchange, navigation, and control, autonomous intelligent vehicles (AIVs) are susceptible to remote hacking in a variety of methods. The communication routes that these vehicles employ to link to other vehicles, infrastructure, and cloud services include Cellular networks, vehicle-to-everything (V2X) technologies, and Wi-Fi. These could all have vulnerabilities that cybercriminals could take advantage of channels to get unauthorized access to the car's critical systems [6] (Table 1).

(ii) Manipulation of Sensors

Table 1 Typical online dangers to self-driving cars and their consequences

Category	An explanation	Actual examples	Possible effects
Hacking remotely	Using wireless connectivity to get unauthorized access to car systems	Hacking of the Jeep Cherokee (2015)	Safety hazards, taking over vehicle operation, and turning off features [7].
Modification of sensors	Disruption of LiDAR, radar, and camera sensors	Tesla's 2016 autopilot scam	Misidentification of obstacles, unpredictable actions, and collisions [8]
Data breaches	Unauthorized access to private information that the car transmits or stores	The 2020 server hack of an electric vehicle manufacturer	Identity theft, privacy violations, and impaired decision-making [9].
DoS attacks	Overloading the car's systems to interfere with regular functions	DDoS assaults on networks connecting vehicles to infrastructure	Vehicle immobilisation, connection loss, and performance deterioration

For precise detection and interpretation of their surroundings, AIV primarily relies on a complex variety of sensors such as LiDAR, ultrasonic sensors, radar, and cameras. The crucial information required for decision-making, obstacle identification, and navigation is supplied by these sensors. AIVs are susceptible to sensor manipulation, including spoofing and jamming, nevertheless, because of their dependence on these sensors. Sensor faking is the practice of introducing erroneous data into sensor systems, whereas jamming interferes with the sensors' regular operation by flooding them with signals or noise [10, 11].

Attackers can generate false scenarios by manipulating sensors, which causes the vehicle's control systems to interpret the surroundings incorrectly. For example, crashes may occur if real obstacles are concealed. But projecting fictitious hazards into the path of the vehicle can cause abrupt and needless braking. These kinds of manipulations may lead to unpredictable driving patterns, which could seriously endanger the safety of other drivers, pedestrians, and passengers. Furthermore, because sensor jamming prevents the vehicle from receiving accurate sensory information, it may become momentarily blind to its surroundings, putting it in potentially hazardous scenarios [12].

(iii) Breach of Data

Autonomous Intelligent Vehicles (AIVs) collect and handle a wide range of data, including sensitive data such as real-time location, driving behaviors, and passenger personal information. The operation and development of AIV systems depend on this comprehensive data collection, which enables them to enhance user experience, optimize routes, and navigate accurately. However, because of the sheer volume and sensitivity of this data, AIVs are prime targets for data breaches [13].

There are several ways that AIV-related data breaches might happen, including gaining unauthorized access to onboard data storage, intercepting communication networks, or taking advantage of flaws in cloud services that facilitate vehicle operations. Cybercriminals can use methods including breaking into the car's internal systems, intercepting data sent between the car and outside parties, or taking advantage of lax security in linked infrastructure [14].

In AIVs, data breaches have serious and varied repercussions. Exposure of sensitive data may result in major privacy violations, including real-time location tracking and the unauthorized sharing of passengers' personal information. This may result in stalking, identity theft, and other misuses of personal data. Beyond privacy issues, corrupted data integrity is a serious threat to AIV's dependability and security. AIVs are susceptible to making bad decisions that result in mishaps or malfunctions if the data that informs their decision-making processes is changed or contaminated.

Preventing data breaches requires putting in place strong cybersecurity safeguards that guarantee data availability, confidentiality, and integrity. The following are a few of these actions:

1. Encryption: Use strong encryption techniques for data at rest and in transit to prevent unauthorized access [15]
2. Authentication and Access Control: Strict authentication and access control protocols must be implemented to guarantee that sensitive data is accessible to only authorised individuals and systems [16].
3. Monitoring network traffic with sophisticated intrusion detection systems (IDSs) to spot suspicious activity that could indicate a breach attempt [17]
4. To find and fix any possible flaws in the antiviral systems, conduct routine security audits and penetration tests [18].
5. Data Anonymization: Personal data should be anonymized whenever possible to lessen the possible impact of a data breach on individual privacy.

(iv) Denial of Service (DoS) Attacks

AIVs' processing and communication capabilities are the subject of assaults like denial of service (DoS), which pose a severe danger to their regular operation. Attackers try to flood the network with traffic to overwhelm the car's systems. or by taking advantage of certain vulnerabilities to bring down or malfunction of vital components. The AIV's capacity to transmit and receive information is hampered by this, which may have dire repercussions right away [19].

There are several DoS attacks that can target AIVs. Attackers may use network flooding techniques to overload the communication channels, making it difficult or impossible for the automobile to send and receive the data it requires. This can lead to a loss of connectivity with critical external systems, such traffic management servers or other vehicles, in a vehicle-to-everything (V2X) context. Another tactic that could halt essential functions and make the vehicle unresponsive is to take advantage of vulnerabilities in the AIV's operating systems or applications [19, 20].

Several tactics can be used to reduce the likelihood of DoS attacks, such as the following:

(i) By setting up backup computational and communication channels, the car can function even in the event that one conduit fails. If an attack occurs, failover solutions can swiftly switch to backup systems.

(ii) By detecting and halting undesired traffic patterns, rate limitation and traffic filtering are two techniques that can reduce network overload [21].

(iii) System for Detecting Anomalies: DoS assaults can be detected and stopped before they inflict serious harm. Employing anomaly detection systems, which in real time monitor network traffic and system behaviour. By employing machine learning techniques, these systems can identify odd patterns that can point to an attack [22].

(iv) Resource Management: By putting effective resource management procedures in place, important systems may be guaranteed the bandwidth and processing power they need to operate properly, even in the face of an assault [23].

(v) Frequent Security Updates: The danger of DoS attack exploitation can be decreased by making sure that all software components are patched and updated on a frequent basis to address known vulnerabilities [24].

Current Countermeasures

Many innovative cybersecurity techniques have been created and put into practice to tackle the various and constantly changing cybersecurity threats that AIVs face.

The purpose of these safeguards is to keep the vehicle's critical systems and data safe against unauthorized access, change, and interference (Table 2) [25].

Table 2 Measures to combat cybersecurity threats in AIV

Countermeasure	Description	Benefits
Intrusion detection systems	Keeping an eye on network traffic for potentially harmful activity	Threat detection and anomaly identification in real time [26]
Encryption	Protecting data while it's in motion and at rest	Safeguards the confidentiality and integrity of data [27]
Regular updates	Firmware and software updates via OTA	Fixed vulnerabilities and improved functionality (Qureshi et al. 2022).
Authentication protocols	Making sure that car systems are only accessible by authorized	Stops unwanted access and protects communication [28].

Table 3 A summary table comparing blockchain-based mitigating techniques and cybersecurity threats

S/N	Cybersecurity threat	Blockchain-based mitigation strategy
1	51% attack	Use either delegated proof of stake or proof of stake
2	Phishing	Adopt two-factor authentication (2FA) and provide users with education
3	Smart contract vulnerabilities	Perform comprehensive smart contract testing and audits
4	DDOS attacks	Put in place a decentralized infrastructure and rate limitation
5	Theft of keys	Make use of hardware wallets and multi-signature wallets
6	Code exploitation	Perform routine testing and code audits
7	Data tampering	Employ an unchangeable Blockchain ledger to stop data manipulation
8	Identity theft	Put in place decentralized identity management systems
9	Man-in-the-Middle attacks	Employ secure communication methods and end-to-end encryption
10	Single point of failure	Decentralised architecture should be used to get rid of single points of failure

3 Cybersecurity Threats Versus Blockchain-Based mitigation strategies

By evaluating cybersecurity data and developing the required processing tools, cybersecurity attacks can be avoided in addition to a set of basic functional requirements and knowledge about risks, threats, or vulnerabilities. Risks can be successfully decreased by employing blockchain-based mitigation techniques. The cybersecurity risks and several blockchain-based countermeasures are listed in Table 3.

4 Privacy Issues in Connected Vehicle Systems

AIV considers privacy to be a crucial component. Autonomous intelligent vehicles raise Some privacy and security issues that need to be resolved in order to guarantee the responsible and secure development and use of this technology. These issues include:

(i) Privacy of Identity

The confidentiality of a participant's real-world identity, as well as the identities of drivers, passengers, pedestrians, and travelers, are all included in AIV's concept of identity privacy. Their first and surname names, driver's license number, vehicle registration number, etc., might all be used to indicate this. Instead of assigning real-world identities to cars that are a part of Vehicular Ad hoc Networks (VANET) systems, pseudo IDs, commonly referred to as pseudoonyms, have been suggested in current Vehicular Ad hoc Networks (VANET) research. It has been demonstrated that pseudonyms are an effective way to secure the connection between the sender's identity and message broadcasts in Vehicular Ad hoc Networks (VANETs) that provide safety information, like vehicle position [29].

(ii) Behavioral Privacy

There is a significant chance that people's behavioural privacy could be violated because AIV contains a wealth of precise information on its users, including location data, financial information, and user habits [30]. have investigated different privacy classification approaches. Behavioural privacy, as used in AIV, refers to the privacy of information that characterizes different facets of a group or individuals and their behaviour inside AIV. A system must be able to anonymize and shield gathered user data from disclosure in addition to concealing typical AIV user behaviour patterns, in order to preserve behavioural privacy.

Protecting the privacy of AIV users' activity is essential to preventing hackers from following and making assumptions about particular users within the system. Individual travellers' movement patterns are also recorded in the system as part of AIV's collection of data on passenger routing patterns to make the routes safer and more effective; this data can be analyzed to draw conclusions about the behaviours of specific persons. Individual travellers' origin and destination points, for instance, may give rise to privacy concerns since they may allow a malevolent actor to deduce a traveler's place of employment or domicile.

(iii) Location Privacy

According to the definition of privacy provided by [30], location privacy in AIV would fall under the category of privacy of location and space, which refers to the user's ability to move freely across the system without fear of their location information being compromised. While precise location information is necessary for AIV to provide location-aware services, it can also be misused to infringe on people's privacy. With this GPS information, attackers can conduct targeted attacks. GPS-based navigation systems have a hard time operating while maintaining user privacy about their whereabouts. Thus, it's crucial to find a balance between protecting users' privacy about their location and providing them with accurate and useful services.

5 Security Risks in Vehicle-to-Vehicle (V2V) and Vehicle-to-Infrastructural (V2I) Communication

Since the creation and improvement of software-driven autonomous driving and vehicle-to-vehicle (V2V) communication technologies, the automobile sector has seen tremendous progress in recent years. By improving safety, effectiveness, and connection amongst vehicle systems, these advances have revolutionised the transportation industry (11). Additionally, as a result of this development, smart cars that are safer and more dependable than ever before have emerged. Numerous security concerns are included in cybersecurity in connected and autonomous vehicle (CAV) environments [19].

For the automobile industry and public safety, the advent of V2V, V2I, and autonomous intelligent vehicles (AIVs) has brought up new cybersecurity challenges, including sensor manipulation, denial-of-service (DoS) attacks, GPS spoofing, man-in-the-middle attacks, and unauthorized data access. AIV's constant data transmission raises cybersecurity risks, such as concerns about data privacy. AIV systems are also vulnerable to attacks like hacking, which jeopardise passenger safety by compromising vital functions like steering and braking, and spoofing, which can interfere with traffic flow [31]. To navigate and make judgments in real time, AIVs that use vehicle-to-cloud (V2C) technology depend on a variety of intricate sensors, communication tools, and artificial intelligence [25, 32].

Finding cybersecurity vulnerabilities in Vehicle-to-Everything (V2X) interfaces, including V2V and V2I interactions, requires an understanding of how connected and autonomous cars (CAVs) communicate, as illustrated in Fig. 1. Analyzing how different cyberthreats can enter the AIV ecosystem through exposed nodes and channels is possible with this network method. Despite being a component of the infrastructure, Fig. 1 highlights the distinct functions that roadside units (RSUs), base stations, and cell towers play in wireless communication within the V2X ecosystem by depicting them independently.

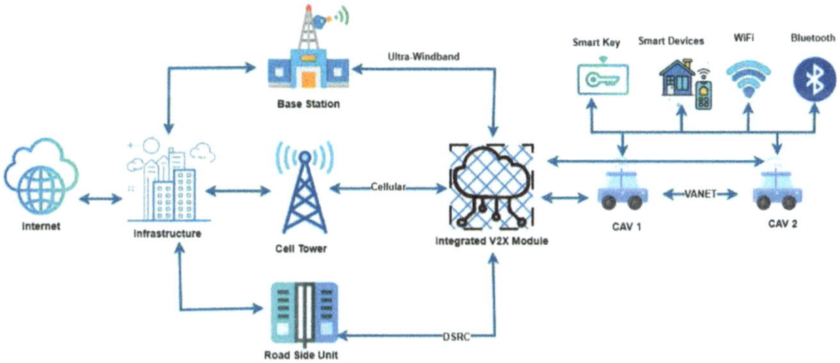

Fig. 1 The CAVs' communication platform, which serves as the foundation for detecting vulnerabilities in cybersecurity [33]

6 Blockchain as a Security Layer

Blockchain is considered to be one of the most popular technologies in recent years due to its inherent ability to become a trustworthy source. Numerous applications have made extensive use of blockchain technology, including cryptocurrency or electronic cash, financial transactions, healthcare, insurance, IoT, manufacturing, AIV, education, etc., because of its decentralized nature and reliability. By documenting every action taken by every entity inside the blockchain network, a blockchain framework is suggested for the security and transparency of the linked cars and users. Its purpose is to guarantee confidentiality and openness between drivers and clients.

Seven separate layers make up the strong security foundation upon which blockchain technology is based. Every layer is essential to guaranteeing the availability, confidentiality, and integrity of data on the blockchain. These layers are as follows:

(i) Physical Layer

In reality, the bits are carried by the physical layer. The majority of the devices in this layer are Internet of Things (IoT) devices, which serve as blockchain nodes and are connected to the internet. Smart contracts transform traditional contractual stipulations into embedded hardware and software, allowing the blockchain to become decentralized [34]. To connect to a blockchain-based system, an IoT device needs to interface with smart contracts, complete a digital signature, and perform other authentication steps [35]. The inability of the blockchain to communicate with IoT devices puts device security and data privacy at risk.

(ii) Data Layer

The blockchain's data structure is represented by this layer. A combination of valid transactions made up of a block header and a block body is called a block in a shared ledger. A header is composed of various components, including the block version, timestamp, Nonce, bits, asymmetric encryption, and Merkle root hash. In the block body, there are numerous transactions. The development of a modified version of the transaction signature or hash collisions are linked to security concerns, as will be covered later.

(iii) Network Layer

Due to its ability to facilitate communication between nodes and the spread of blocks and transactions, the network layer is essential to the operation of blockchain systems. The network is vulnerable to a number of assaults, including Sybil, Eclipse, DDoS, and routing attacks, due to its open and decentralized structure.

(iv) Consensus Layer

There are several methods used in the consensus layer to ensure that every node in the decentralized network agrees on the data validity of newly created blocks. Consensus-based agreements are used by blockchain networks to validate and successfully add new blocks to the chain [19, 36]. This consensus across nodes is based on the hash rate and is accomplished through proof of work. A significant amount of Bitcoin processing power is used by miners to compete to solve a cryptographic challenge in order to validate transactions and produce a new block. The decentralization and security of the network are affected by this process [36].

(v) Incentive Layer

The blockchain system requires several trustworthy nodes to confirm and validate every transaction to maintain security and decentralization. To encourage nodes to contribute to preserving the system's security, incentive mechanisms are needed. In order to guarantee that the majority of the network is honest, the incentive mechanism is essential to the blockchain system.

(vi) Contract Layer

On blockchain platforms like Ethereum, smart contracts live and operate at the contract layer. These self-executing programs facilitate transactions without the need for middlemen, maintain digital assets, and enforce agreements. However, attackers may use smart contract flaws to alter the state of the contract, steal money, or interfere with contract behavior.

(vii) Application Layer

This layer includes programmable cash, programmable banking, and programmable society, among other application scenario types. Malicious exchanges and service providers, malware, design, configuration, and internal and external attackers are just a few of the many dangers that might arise within this layer. The exchange operator (or malevolent operator) has complete access to the funds on their systems, which puts users' assets at risk [37] (Table 4).

7 Blockchain-Based Frameworks

In this article, we provide a brief overview of blockchain frameworks for AIV data privacy, liability assignment, and secure platooning.

(i) Combining Blockchain and Aggregated Zero-Knowledge Proof (ZKP) for Dynamic Truck Platooning and Authentication

Maintaining a record of the authentication events for auditing purposes and authenticating AIVs entering dynamic groups without disclosing their identities are the goals. For this paradigm, a mix of blockchain and aggregated zero-knowledge proofs was suggested in [38]. The network, application, and data storage layers make up the suggested framework. It can be used in situations where privacy and verifiability are

Table 4 Mapping of Blockchain Layer to AIV Threat Mitigated

S/N	Blockchain layer	AIV threat mitigated
1	Physical layer	Node tampering, node spoofing
2	Data layer	Data manipulation, false log attacks, and disputes over liability after collisions between AIVs
3	Network layer	Message tampering, single-point
4	Consensus layer	Spoofing, sybil attacks
5	Incentive layer	Spoofing, false reporting
6	Contract layer	Relay attacks, reward manipulation
7	Application layer	Malicious policy changes, malicious exchanges among service providers, and latent manipulation of incentives by nodes

important considerations. It is still difficult to generate and validate ZKP on limited devices, such as sensors.

(ii) A Reverse-Traceability Framework for Determining Self-Driving Car Liability

The purpose of this framework is to track down and identify the likely individual or individuals and/or program that caused an accident [2]. Vehicle event logs are connected to on-chain records to do this. The application, data storage, and physical layers make up the suggested framework. When an accident occurs, it is incredibly helpful for adjudication and litigation in the real world. However, cross-industry agreements are necessary for smart contracts to have legal force behind them.

(iii) Blockchain-enabled Data Security and Vehicular Consensus (VBCA) for Vehicular Networks

The goal of this architecture is to guarantee safe message exchange in AIV systems and vehicular area networks (VANETs) [39]. Message integrity, defence against specific kinds of assaults, and verification of crucial V2X communications are its primary concerns. Consensus, data, and network layers make up the structure. It provides a solution to bridge the performance gaps between VANET timing requirements and blockchain. Scaling it to the local or regional level is still an unresolved issue.

8 Review of Related Works

The paradigm change that is occurring in the automobile sector is the result of a wave of modernization and technical advancement. Level 2 and level 3 AVs will account by for 92 and 8% of the market share, respectively, by 2030 [40]. Projections indicate that the AV market will increase by 39.47%, from USD 54.23 billion in 2019 to USD 75.6 billion in 2026, and USD 87 billion by 2030. By 2035, it's expected that 25% of all automobile sales will be self-driving. Although North America is anticipated to dominate the AV market, Pütz et al. (2019) predict that Europe will have the fastest growth rate in the AV market between 2019 and 2026, at 42.6%.

Other transportation-related problems, such as traffic and accidents, are also anticipated to be resolved by autonomous intelligent cars. For example, human error accounts for 93% of accidents, particularly crashes. By lowering human involvement in driving, autonomous intelligent cars will lower these numbers and, in turn, human faults including speeding, distracted driving, and drunk driving. Semi-autonomous intelligent cars have already shown this effect on reducing collisions.

A study by the Insurance Institute of Highway Safety found that partially autonomous features, including forward collision avoidance, side view assistance, and lane departure warning, reduced road accidents, fatalities, and collisions by at least 33% [41, 42]. agree that adding AI to AIVs will increase their ability to address some of the problems with conventional cars, including poor efficiency, traffic congestion, pollution, road safety, and the lack of independence for people with impairments.

Nonetheless, there is a chance for AIVs to work together to manage traffic in a way that will probably reduce congestion over time. Among the components is the platooning of automobiles, in which changing routes could enhance traffic flow by more accurately coordinating the speeds and distances between vehicles. Better connections for AIVs, such as smart traffic signals and dedicated AIV lanes to lessen congestion points, are examples of possible changes. Adoption of AIVs is anticipated to bring in long-term prospects for improved traffic systems, even though their use may initially cause minor congestion.

The concerns about privacy posed by data collection in autonomous driving systems are causing concern. Location information, when paired with other types of personal information, can reveal important details about a person's traits, including their financial standing, work status, sexual orientation, and religious affiliations. Significant cybersecurity flaws, including remote hacking and data privacy issues, might affect connected cars. The goal of remote hacking is to obtain unauthorised access to the car's systems. They can directly affect or have an impact on vital processes like steering and braking.

[43] presented a method that combined BC and AI in 2023 [43, 44] introduced a BC system that uses deep learning to protect the privacy of futuristic cars, [45] developed a distributed approach to protect passengers' privacy while they are driving. Using Blockchain to enhance the AIV network's secrecy and integrity in 2022, [7] wrote a

paper about the prospect and difficulties of integrating blockchain technology with ACVs.

A BC framework was introduced to Rathee in 2019 with the goal of improving the security of autonomous and networked cars [46]. A poll in 2022 regarding BC and its possible use in smart settings to improve security and privacy [46, 47]. has developed a technique to guarantee secure vehicle communication using BC [47]. Smys discovered that adding a BC-based architectural framework could improve smart car security

9 Method

The systematic procedure that was used to examine cybersecurity problems and fixes pertaining to autonomous intelligent vehicles (AIVs) is detailed in this section. In order to ensure a comprehensive and impartial assessment, the approach includes the entire review process, databases, inclusion and exclusion criteria, and search term selection. Numerous search terms, such as cybersecurity threats, privacy concerns in connected vehicle systems, security risks in vehicle-to-vehicle communications, autonomous intelligent vehicles, and vehicle-to-infrastructure communications, were used to cover the wide range of cybersecurity issues in autonomous intelligent vehicles. To guarantee a broad range of sources, literature search was conducted using several significant databases, including the ACM Digital Library, IEEE Xplore, Google Scholar, PubMed, and Science Direct. To guarantee reliability and verifiability of sources, stringent inclusion and exclusion criteria were applied. It was thought that papers released between 2015 and 2023 will cover new developments and dangers.

10 Results

In order to satisfy the study objectives, the book chapter's results will be presented in multiple sections using tables and figures to provide readers with accurate and sufficient information.

11 Discussion

Blockchain technology has the potential to transform a wide range of businesses, including the AIV. The two main problems facing this network of linked devices are security and trust because the AIV ecosystem is growing so rapidly. By depending just on standard practices, the AIV cannot be safeguarded against these problems, such as guaranteeing data integrity and accurate authentication. Therefore, by reducing manipulation and encouraging transparency and decentralization, this book chapter suggests blockchain as the most effective means of enhancing the security and reliability of the Internet of Things.

One of the most important features of blockchain technology is its ability to shield data integrity against unauthorized change or tampering. Since the data cannot be altered or tampered with until all parties agree, once it is recorded on the blockchain and approved by the network parties, it is protected from unauthorized access. This feature can be used to audit the data path, find its source, and verify its legitimacy. This gives AIV networks confidence and establishes an impenetrable environment [2]. Similar to the AIV network, massive amounts of data are produced when billions of devices are connected.

Blockchain uses cryptographic characteristics to keep this data safe and immutable, and it is recorded in a distributed ledger. Furthermore, it ensures that the data is not controlled by a single party, reducing the risk of fraud and encouraging openness and decentralization.

In conclusion, there are several ways that blockchain technology might help the AIV. Security is enhanced by using a decentralized structure. It boosts trust by verifying that the information is authentic, reliable, and unaffected. It eliminates the need for middlemen and enables peer-to-peer data sharing. The flexibility and reliability of distributed consensus are advantageous for AIV systems. It guarantees regulatory compliance by streamlining and optimizing compliance and audit processes. Most people agree that blockchain technology can help address security and trust issues in AIV systems.

Acknowledgments The writers express their gratitude to every editor and reviewer who helped to enhance the publication's substance.

References

1. Bathla, G., Bhadane, K., Singh, R.K., Kumar, R., Aluvalu, R., Krishnamurthi, R., Kumar, A., Thakur, R.N., Basheer, S.: Autonomous vehicles and intelligent automation: applications, challenges, and opportunities. Mob. Inf. Syst. **2022**, 1–9 (2022)
2. Gupta, S., Verma, J.K.: Reverse traceability framework for identifying liability of crashes for self-driving vehicles using blockchains. J. Glob. Inf. Manag. (JGIM). **31**(7), 1–21 (2023)
3. Xu, Q., Li, H.: A review of intrusion detection in vehicular networks. IEEE Trans. Intell. Transp. Syst. **22**(6), 3451–3463 (2021)
4. Kim, S., Shrestha, R.: Security and privacy in intelligent autonomous vehicles. In: Automotive Cyber Security, pp. 35–66. Springer (2020). https://doi.org/10.1007/978-981-15-8052-9
5. Algarni, A., Thayananthan, V.: Autonomous vehicles: the cyber security vulnerabilities and countermeasures for big data communication. Symmetry. **14**(2494) (2022)
6. Malik, S., Sun, W.: Analysis and simulation of cyber attacks against connected and autonomous vehicles. In: Proceedings of the 2020 International Conference on Connected and Autonomous Driving (MetroCAD)pp. 62–70 (2020)
7. Dargahi, T., Ahmadvand, H., Alraja, M.N., Yu, C.M.: Integration of blockchain with connected and autonomous vehicles: vision and challenge. ACM J. Data Inf. Qual. **14**(1), 1–10 (2021)

8. Azam, S., Munir, F., Sheri, A.M., Kim, J., Jeon, M.: System, design and experimental validation of autonomous vehicle in an unconstrained environment. Sensors. **20**(5999) (2020). https://doi.org/10.3390/s200205999
9. Frigerio, A., Vermeulen, B., Goossens, K.G.W.: Automotive architecture topologies: analysis for safety-critical autonomous vehicle applications. IEEE Access. **9**, 62837–62846 (2021)
10. Chen, L., He, Y., Wang, Q., Pan, W., Ming, Z.: Joint optimization of sensing, decision-making and motion-controlling for autonomous vehicles: a deep reinforcement learning approach. IEEE Trans. Veh. Technol. **71**, 4642–4654 (2022)
11. Wang, H., Liu, Y.: Blockchain for autonomous vehicles: A comprehensive survey. IEEE Trans. Intell. Transp. Syst. **22**(6), 3238–3256 (2021)
12. El-Rewini, Z., Sadatsharan, K., Sugunaraj, N., Selvaraj, D.F., Plathottam, S.J., Ranganathan, P.: Cybersecurity attacks in vehicular sensors. IEEE Sensors J. **20**, 13752–13767 (2020)
13. Liu, N., Nikitas, A., Parkinson, S.: Exploring expert perceptions about the cybersecurity and privacy of connected and autonomous vehicles: a thematic analysis approach. Transport. Res. F: Traffic Psychol. Behav. **75**, 66–86 (2020)
14. Katrakazas, C., Theofilatos, A., Papastefanatos, G., Härri, J., Antoniou, C.: Cyber security and its impact on CAV safety: overview, policy needs and challenges. Adv. Transp. Policy Plan. **5**, 73–94 (2020)
15. Benyahya, M., Collen, A., Nijdam, N.A.: Analyses on standards and regulations for connected and automated vehicles: identifying the certifications roadmap. Transp. Eng. **14**, 100205 (2023)
16. Maeng, K., Kim, W., Cho, Y.: Consumers' attitudes toward information security threats against connected and autonomous vehicles. Telematics Inform. **63**, 101646 (2021)
17. Karopoulos, G., Kambourakis, G., Chatzoglou, E., Hernández-Ramos, J.L., Kouliaridis, V.: Demystifying in-vehicle intrusion detection systems: a survey of surveys and a meta-taxonomy. Electronics. **11**, 1072 (2022)
18. Mateo Sanguino, T.D.J., Lozano Domínguez, J.M., de Carvalho Baptista, P.: Cyber security certification and auditing of automotive industry. Adv. Transp. Policy Plan. **5**, 95–124 (2020)
19. Sun, W., Malik, S.: Cyber-physical security for connected and autonomous vehicles: a survey. Electronics. **8**(8), 864 (2019)
20. Alsaade, F.W., Al-Adhaileh, M.H.: Cyber attack detection for self-driving vehicle networks using deep autoencoder algorithms. Sensors. **23**(4086) (2023)
21. Basiri, M.H., Azad, N.L., Fischmeister, S.: Secure dynamic nonlinear heterogeneous vehicle platooning: denial-of-service cyber-attack case. In: Security in Cyber-Physical Systems, pp. 287–315. Springer (2021)
22. Khadka, A., Karypidis, P., Lytos, A., Efstathopoulos, G.: A benchmarking framework for cyber-attacks on autonomous vehicles. Transp. Res. Procedia. **52**, 323–330 (2021)
23. Nayak, B.P., Hota, L., Kumar, A., Turuk, A.K., Chong, P.H.J.: Autonomous vehicles: resource allocation, security, and data privacy. IEEE Trans. Green Commun. Netw. **6**, 117–131 (2022)
24. Limbasiya, T., Teng, K.Z., Chattopadhyay, S., Zhou, J.: A systematic survey of attack detection and prevention in connected and autonomous vehicles. Veh. Commun. **37**, 100515 (2022)
25. Chow, M.C., Ma, M., Pan, Z.: Attack models and countermeasures for autonomous vehicles. In: Intelligent Technologies for Internet of Vehicles, pp. 375–401. Springer (2021)
26. Kim, K., Kim, J.S., Jeong, S., Park, J.-H., Kim, H.K.: Cybersecurity for autonomous vehicles: review of attacks and defense. Comput. Secur. **103**, 102150 (2021)
27. Karim, H., Rawat, D.B.: TollsOnly please—Homomorphic encryption for toll transponder privacy in internet of vehicles. IEEE Internet Things J. **9**, 2627–2636 (2022)
28. Malik, S., Bandi, P., Sun, W. An experimental study of denial of service attack against platoon of smart vehicles. In Proceedings of the 2021 Fourth International Conference on Connected and Autonomous Driving (MetroCAD) pp. 23–30 (2021)
29. Nkenyereye, L., Tama, B.A., Park, Y., Rhee, K.H.: A fine-grained privacy preserving protocol over attribute based access control for VANETs. J. Wirel. Mob. Netw. Ubiquitous Comput. Dependable Appl. **6**(2), 98–112 (2015)
30. Finn, R.L., Wright, D., Friedewald, M.: Seven types of privacy. In: European Data Protection: Coming of Age, pp. 3–32. Springer (2013)

31. Parkinson, S., Ward, P., Wilson, K., Miller, J.: Cyber threats facing autonomous and connected vehicles: future challenges. IEEE Trans. Intell. Transp. Syst. **18**(11), 2898–2915 (2017)
32. Chowdhury, I., Hasan, R. Security analysis of connected autonomous vehicles (CAVs): challenges, issues, defenses, and open problems. In: Proceedings of the IEEE World Forum Public Safety Technology (WFPST). IEEE (2024)
33. Arachchige, K.G., Alkaabi, G., Murtaza, M., Haq, Q.E.U., Abuvalkishik, Z., Lee, C.-C.: Threat landscape and integrated Cybersecurity framework for V2V and autonomous electric vehicles. World Electr. Veh. J. **16**, 469 (2025). https://doi.org/10.3390/wevj16080469
34. Choo, K.-K.R., Dehghantanha, A., Parizi, R.M.: Blockchain Cybersecurity, Trust and Privacy. Springer International Publishing (2020)
35. Yang, J., Bi, H., Liang, Z., Zhou, H., Yang, H. J.: A survey on blockchain: architecture, applications, challenges, and future trends. In: Proceedings of the International Conference on big data and Computing, pp. 19–24. ACM (2020)
36. Beikverdi, A., Song, J. S.: Trend of centralization in bitcoin's distributed network. In: Proceedings of the 2015 IEEE/ACIS 16th International Conference on Software Engineering, Artificial Intelligence, Networking and Parallel/Distributed Computing (SNPD*)* pp. 1–6. IEEE (2015)
37. Annessi, R., Fast, E.: Improving security for users of decentralized exchanges through multiparty computation. In: Proceedings of the 2021 IEEE International Conference on Blockchain (Blockchain) pp. 229–236. IEEE (2021)
38. Li, W., et al.: Aggregated zero-knowledge proof and blockchain-empowered authentication for autonomous truck platooning. IEEE Trans. Intell. Transp. Syst. **24**(9), 9309–9323 (2023)
39. Sehar, N.U., Khalid, O., Khan, I.A., et al.: Blockchain enabled data security in vehicular networks. Sci. Rep. **13**, 4412 (2023). https://doi.org/10.1038/s41598-023-31442
40. Grey B. (2022) Top 30 self-driving technology and car companies. Retrieved from 23 10 2022
41. Anderson, J.M., Nidhi, K., Stanley, K.D., Sorensen, P., Samaras, C., Oluwatola, O.A.: Autonomous Vehicle Technology: a Guide for Policymakers. Rand Corporation (2014)
42. Karnati, A., Mehta, D.: Artificial intelligence in self-driving cars: applications, implications and challenges. Ushus J. Bus. Manag. **21**(1), 1–28 (2022)
43. Benyahya, G., Hameurlaine, A., Germanos, G., Kolokotronis, N., Shiaeles, S.: Autonomous vehicles security: challenges and solutions using blockchain and artificial intelligence. IEEE Trans. Intell. Transp. Syst. **24**, 3614–3637 (2023)
44. Nair, M.M., Nair, S.M., Tyagi, A.K.: Deep learning based blockchain solution for preserving privacy in future vehicles. Int. J. Hybrid Intell. Syst. **16**(4), 223–236 (2020)
45. Dorri, A., Steger, M., Kanhere, S.S., Jurdak, R.: Blockchain: a distributed solution to automotive security and privacy. IEEE Commun. Mag. **55**(12), 119–125 (2017)
46. Ebrahim, M., Hafid, A., Elie, E. Blockchain as privacy and security solution for smart environments: A survey. https://arxiv.org/abs/2203.08901 (2022)
47. Koduri, R., Nandyala, S., Manalikandy, M.: Secure Vehicular Communication Using Blockchain Technology (No. 2020-01-0744). SAE Technical Paper (2020)

Smart Contracts for Secure Autonomous Vehicle Operation

Esmot Ara Tuli, Mohtasin Golam, and Dong-Seong Kim

Abstract Blockchain is increasingly being explored for trustworthy coordination, data integrity, and automation across distributed autonomous vehicles (AVs). A smart contract is the programmable substrate for applying logic, access control, governance, and other operations in a blockchain environment. To automate blockchain-based operations in an AV system, smart contract is crucial. This book chapter explores the application landscape for smart contract in AVs and outlines the corresponding security requirements. Although smart contract is considered here for security assurance, it is important to first ensure the security of the smart contracts itself. Therefore, vulnerability identification, and security assurance for smart contracts are discussed in this chapter. Moreover, security challenges posed by advanced technologies, such as quantum computing, in smart contracts for AV operations are also highlighted in this chapter. Finally, how quantum computing can be applied to ensure security in smart contracts is also mentioned, with a case study showing quantum machine learning-based vulnerability detection in smart contracts for AV operations.

Keywords Autonomous vehicles · Blockchain · Smart contract · Quantum computing

Esmot Ara Tuli and Mohtasin Golam These authors contributed equally to this work.

E. A. Tuli
ICT Convergence Research Center,Kumoh National Institute of Technology,Gumi, South Korea

M. Golam · D.-S. Kim (✉)
NSLab Inc. ICT Convergence Research Center,Kumoh National Institute of Technology,Gumi, South Korea
e-mail: dskim@kumoh.ac.kr

M. Golam
e-mail: gmoh248@kumoh.ac.kr

S. O. Ajakwe and D.-S. Kim (eds.), *Blockchain for Security of Autonomous Intelligent Vehicles*, Studies in Big Data 194,
https://doi.org/10.1007/978-3-032-20851-4_7

1 Introduction

Autonomous vehicles (AVs) are advancing at a rapid pace, with the help of concurrent improvements in several enabling technologies. Key among these are artificial intelligence (AI), smart sensors, and specialized vehicular networking systems (V2X, V2E, and others). These technologies are essential for the functionality of AVs, as they allow vehicles to operate with reduced reliance on human control and manual input. As AVs evolve, the interconnectivity and complexity of their systems increase significantly, encompassing not only the vehicles themselves but also their underlying network infrastructures and architectural designs.

AV automation levels are classified into five different labels according to the standards defined by SAE J3016 [1]. From Level 3 onwards, AVs possess the capability to make decisions based on situational awareness, such as detecting obstacles or adapting to changing environments, without direct human intervention [2]. This progression in automation introduces new challenges, particularly in the area of cybersecurity. A seemingly minor security breach could have a profound impact on both the AV and its broader operational ecosystem. To ensure robust security in AVs, a range of protective measures are essential. These include secure authentication protocols, the aggregation of machine learning models, decentralized processing and aggregation of sensor data, and the maintenance of vehicular network security. In recent years, blockchain technology has emerged as a promising solution to address these challenges. Blockchain, a decentralized and immutable ledger system, offers enhanced transparency and tamper-proof data management, making it a popular choice among researchers in the field. Additionally, to automate blockchain transactions and ensure that security protocols are executed as needed, smart contracts are employed. In the context of AVs, smart contracts enable the automation of blockchain-based processes, providing a seamless, secure environment for both data handling and transaction management.

2 Role of Smart Contracts in Self-Driving Vehicle and Autonomous Vehicles

The integration of smart contracts within autonomous intelligent vehicles (AIVs) represents a paradigmatic transformation in vehicular computing architectures, fundamentally reshaping how self-driving vehicles interact with their operational ecosystem. These programmable, self-executing digital agreements serve as the cornerstone of decentralized vehicular networks, enabling sophisticated automation capabilities while maintaining security, transparency, and trustless interactions between vehicles, infrastructure components, and service providers [3].

2.1 Foundational Architecture and Blockchain Intelligence Integration

The implementation of smart contracts in autonomous vehicles relies heavily on the concept of blockchain intelligence, which synthesizes blockchain technology with machine learning algorithms to create a distributed computational framework [4, 5]. This hybrid architecture goes beyond traditional centralized vehicular systems. It pushes machine learning capabilities from centralized servers to distributed network nodes. This creates new pathways for secure, real-time decision-making in autonomous vehicle environments [6]. The blockchain intelligence framework addresses fundamental challenges inherent in Internet of Vehicles (IoV) systems, particularly concerning decentralization, cybersecurity, and data privacy preservation in high-mobility environments [7, 8]. Smart contracts within this architecture function as autonomous computational agents deployed on distributed ledger systems, typically implemented on Ethereum Virtual Machine (EVM) platforms installed on roadside units (RSUs) due to their superior processing capabilities compared to in-vehicle edge devices. These contracts maintain persistent execution states across the vehicular network, enabling continuous operation even as vehicles traverse different geographical regions and network segments [3].

2.2 Transactional Automation and Financial Integration

The most immediate application of smart contracts in autonomous vehicles manifests through sophisticated transactional automation frameworks that eliminate traditional intermediary dependencies. These self-executing agreements facilitate seamless payment mechanisms for critical vehicular services, including electric vehicle charging, automated toll collection, parking fee settlement, and even in-vehicle commerce transactions [9]. The automated transaction framework operates through a multi-stage process encompassing merchandise registration, secure cryptographic storage management, payment validation protocols, fund transfer authorization, and access control mechanisms. Recent implementations demonstrate remarkable efficiency improvements, with elliptic curve cryptosystem (ECC) encryption protocols achieving up to 81% performance enhancement while maintaining minimal computational overhead [10]. These systems replace insecure physical transaction modes with cryptographically secured digital protocols, enabling contactless interactions between vehicles and service providers while ensuring transaction integrity through immutable ledger records. The integration of multi-signature wallet technologies further enhances transactional security by requiring multiple cryptographic signatures to authorize high-value transactions [11]. In autonomous vehicle contexts, multi-signature protocols can be configured to require consensus between vehicle systems, owner credentials, and service provider verification before executing payments, creating robust protection against unauthorized transactions [11].

2.3 Data Sharing and Communication Protocols

Smart contracts establish the foundational infrastructure for secure data exchange within vehicular networks, enabling autonomous vehicles to share critical operational information, including traffic conditions, safety alerts, road hazard warnings, and real-time navigation data [11]. The implementation of blockchain-based communication protocols ensures data integrity and authenticity while maintaining privacy protection through advanced cryptographic techniques. The oracle pattern implementation within smart contracts enables autonomous vehicles to access external data sources, including weather information, traffic management systems, and infrastructure sensor networks [12]. Hardware oracles integrated with Internet of Things (IoT) sensors provide real-time environmental data through radio frequency identification (RFID) systems and various sensor networks, creating comprehensive situational awareness for autonomous vehicle decision-making processes.

2.4 Consensus Mechanisms and Network Governance

The operational integrity of smart contract-based vehicular networks relies on sophisticated consensus mechanisms specifically adapted for high-mobility environments. The Proof-of-Quality-Factor (PoQF) consensus mechanism addresses unique challenges in vehicular networks by incorporating signal-to-interference-noise ratio (SINR) measurements to evaluate communication quality across multi-hop vehicle-to-vehicle (V2V) transmission paths [13, 14]. This mechanism provides moderate security performance while maintaining low communication complexity, making it particularly suitable for large-scale vehicular network deployments. Dynamic Proof-of-Work (dPoW) consensus algorithms implement variable difficulty levels that automatically adjust based on incoming traffic patterns in Social Internet of Vehicles (SIoV) environments [15]. These adaptive mechanisms balance security requirements with performance optimization, preventing network hijacking attacks while maintaining efficient transaction processing capabilities.

2.5 Trust Management and Reputation Systems

Smart contracts facilitate sophisticated trust management frameworks within autonomous vehicle networks through automated reputation scoring systems. These contracts maintain immutable trust records for individual vehicles and infrastructure components based on behavioral analysis, message accuracy validation, and historical interaction patterns [16]. The transparency inherent in blockchain technology ensures that trust evaluations cannot be maliciously manipulated while allowing network participants to access trust information for informed decision-making [17]. The implementation of cryptographic identity verification through physical unclonable functions (PUFs) embedded in vehicle system-on-chip (SoC) components creates unique vehicular fingerprints that enable secure authentication and data provenance tracking [18]. This hardware-based security approach prevents identity spoofing attacks and ensures that data originating from vehicles can be cryptographically verified and traced to its source.

2.6 Predictive Maintenance and Digital Twin Integration

The convergence of smart contracts with Digital Twin technology creates sophisticated predictive maintenance frameworks for autonomous vehicle fleets. The IoV-TwinChain architecture utilizes smart contracts to maintain data integrity and traceability across both physical vehicle systems and their virtual representations, enabling real-time monitoring, simulation, and predictive analysis of vehicle operating conditions [19]. This integration optimizes maintenance scheduling algorithms and enables preemptive intervention to address potential mechanical or software issues before they manifest as operational failures. Smart contracts automate the correlation between sensor data from physical vehicles and their digital twin counterparts, triggering maintenance notifications, parts ordering, and service scheduling based on predictive analytics algorithms [20]. This automated approach significantly reduces operational downtime while optimizing resource allocation across vehicle fleets.

2.7 Incentive Mechanisms and Cooperative Behavior

The implementation of blockchain-based incentive systems through smart contracts encourages cooperative behavior among autonomous vehicles within IoV networks. These contracts automatically distribute rewards to vehicles that participate in collaborative tasks such as traffic event reporting, route optimization suggestions, or selection of less congested charging stations [21]. The transparent and immutable nature of smart contract-based reward systems ensures fair compensation distribution while creating economic incentives for network participation. Smart contracts also facilitate the emergence of decentralized autonomous organizations (DAOs) within vehicular networks, enabling collective decision-making processes for traffic management, resource allocation, and network governance [22]. These organizational structures allow vehicle owners and operators to participate democratically in network management decisions while maintaining decentralized operational control.

2.8 Regulatory Compliance and Automated Verification

Smart contracts provide automated compliance verification mechanisms that ensure autonomous vehicles adhere to regulatory requirements across different jurisdictions [23]. These contracts can automatically verify vehicle licensing status, insurance coverage, safety certifications, and emissions compliance before permitting network participation. The immutable audit trails created by smart contract execution provide transparent regulatory compliance records while reducing the administrative overhead associated with manual verification processes [23]. The implementation of automated compliance frameworks becomes particularly crucial as autonomous vehicles operate across multiple regulatory jurisdictions, each with distinct requirements for vehicle operation, data privacy, and safety standards [24]. Smart contracts enable seamless adaptation to varying regulatory environments while maintaining comprehensive compliance documentation. Through these multifaceted roles, smart contracts establish the technological foundation for secure, efficient, and autonomous operation of intelligent vehicles within increasingly complex transportation ecosystems, enabling the realization of fully autonomous transportation networks while maintaining security, privacy, and regulatory compliance [23].

3 Smart Contract Vulnerabilities and Security Issues in AIV

The deployment of smart contracts within autonomous intelligent vehicles (AIVs) introduces a complex landscape of security vulnerabilities that fundamentally differ from traditional blockchain application contexts due to the safety-critical nature of vehicular operations. Unlike conventional smart contract environments, where vulnerabilities primarily result in financial losses, security failures in autonomous vehicle systems can lead to physical harm, compromising passenger safety, disrupting traffic management systems, and potentially causing catastrophic accidents.

3.1 Traditional Smart Contract Vulnerabilities in AIV Context

The fundamental vulnerabilities that plague general-purpose smart contracts become particularly hazardous when manifested within autonomous vehicle ecosystems. Reentrancy attacks represent one of the most critical threats, where malicious contracts exploit recursive function calls to manipulate vehicle payment systems, charging infrastructure access controls, or emergency service protocols [25]. In autonomous vehicle contexts, such attacks could target automated toll payment functions, electric vehicle charging authorization mechanisms, or even critical safety systems that rely on smart contract validation. The notorious DAO hack of 2016, which exploited reentrancy vulnerabilities resulting in millions of dollars in losses, demonstrates the catastrophic potential when such vulnerabilities affect systems with substantial financial stakes [26]. Integer overflow and underflow vulnerabilities pose severe risks to autonomous vehicle operations, particularly in systems handling critical calculations such as fuel management, route optimization algorithms, or collision avoidance protocols [27]. The BEC token incident of 2018 exemplifies how integer overflow vulnerabilities can be exploited to generate arbitrary values [28]. This scenario could prove fatal if applied to autonomous vehicle safety systems where precise mathematical calculations determine braking distances, acceleration rates, or steering corrections. These vulnerabilities become especially dangerous in high-speed autonomous vehicle operations where computational errors could result in immediate physical consequences.

Timestamp dependence vulnerabilities exploit smart contracts' reliance on block timestamps for critical decision-making processes [29]. In autonomous vehicle applications, attackers could manipulate timing-dependent functions related to traffic light synchronization, collision avoidance protocols, or emergency response coordination systems [30]. Such manipulation could create dangerous scenarios where vehicles make incorrect decisions based on falsified temporal information, potentially leading to traffic accidents or system-wide coordination failures. Access control vulnerabilities in autonomous vehicle smart contracts present unique risks due to the distributed nature of vehicular networks [27]. Inadequate permission management could allow unauthorized entities to modify vehicle behavior parameters, access sensitive location data, or inject malicious commands into vehicle control systems. The Parity Wallet hack of 2017, which resulted in the freezing of over 150 million worth of Ether due to access control logic vulnerabilities, illustrates the potential for devastating consequences when applied to autonomous vehicle fleet management systems [31].

3.2 *Privacy-Preserving Data Sharing in AIV*

Permissionless blockchain operates as a decentralized system, distinct from traditional centralized systems, by removing the need for a central authority to maintain the network or ensure privacy. This shift introduces significant challenges, particularly in privacy protection. Unlike traditional centralized models where a central authority can manage data security, in permissionless blockchain systems, privacy is often not guaranteed by default. The open nature of permissionless blockchains allows for greater transparency, meaning that every transaction conducted on the network is publicly recorded and accessible to anyone with the necessary resources to view them. While this can be an advantage for ensuring trust and security through immutability, it also presents privacy concerns.

In a public blockchain, despite the use of cryptographic measures, privacy leakage remains a possibility. By analyzing transaction patterns and linking user addresses, attackers can track individual transactions. Through the examination of transaction flows, it becomes possible to correlate different addresses with a single user, potentially revealing the identity of the user when cross-referenced with external sources of information. This association between addresses can lead to the exposure of sensitive data, undermining the privacy that blockchain technology purports to offer [32].

The importance of privacy becomes even more important in specific applications of AIVs. For instance, in the case of energy trading within electric AIVs, privacy is essential to ensure that the identities of participants remain protected while facilitating transactions. Without privacy, users may be reluctant to engage in such systems, fearing that their behavior and transactions could be tracked and analyzed. Similarly, in distributed deep learning models, such as those used for vehicle navigation or traffic management, ensuring that sensitive data, including user identities, does not leak is crucial to maintaining both security and privacy. Moreover, in systems that rely on autonomous decision-making, such as when forwarding announcements about the vehicle's status or intentions, revealing too much about the vehicle's owner or location could expose the system to malicious attacks or exploitation.

AIV requires vehicles to constantly communicate everything around V2X are depends on complex cloud and edge computing, so ensuring privacy is a massive, active research challenge. This is the reason researchers are researching on different types of privacy-preserving concepts for example, zero-knowledge proofs (ZKPs), differential privacy, homomorphic encryption, and others. ZKPs enable a prover to convince a verifier of the validity of a statement without revealing underlying secrets, aligning with a zero-trust design "never trust, always verify". In the post-quantum scenario, ZKPs are considered promising candidates for long-term privacy and integrity in applications such as privacy-preserving message forwarding, anonymous authentication for cooperative perception, and spectrum-sharing compliance attestation. Differential privacy allows personalized privacy among various users. In AIV, differential privacy could be used for individual driving behavior and trajectory analysis, traffic densities, map updates and others. Homomorphic encryption goes one step further by letting servers do math directly on encrypted data, so fleets can compute statistics or run simple models without ever seeing raw information.

3.3 Vehicular Network-Specific Security Challenges

Autonomous vehicles operating within Internet of Vehicles (IoV) environments face additional security challenges that compound traditional smart contract vulnerabilities. Routing attacks, including blackhole and greyhole attacks, specifically target Vehicular Ad Hoc Networks (VANETs) by exploiting the distributed routing mechanisms inherent in vehicular communication systems [33]. Blackhole attacks involve malicious nodes claiming optimal routes while discarding received packets, effectively creating communication dead zones that could isolate autonomous vehicles from critical infrastructure communications [33]. Greyhole attacks present even more insidious threats by selectively dropping packets, making detection significantly more challenging while still disrupting critical communications [33].

Sybil attacks represent another serious threat to autonomous vehicle networks, where malicious actors create multiple false identities to manipulate consensus mechanisms or flood networks with false information [34]. In the context of smart contract-based location verification systems, Sybil attacks can undermine the integrity of proof-of-location protocols that autonomous vehicles rely upon for navigation and coordination [35]. Recent research demonstrates that Sybil attack mitigation techniques using computational puzzles and smart contracts can achieve false location registration probabilities below 10% [36]. Still, the computational overhead and timing constraints required for such protection mechanisms may not be suitable for high-mobility vehicular environments.

Denial of Service (DoS) attacks targeting autonomous vehicle smart contracts present critical safety risks by potentially rendering essential vehicle functions inoperative [35]. Research on Vehicle-Based Secure Blockchain Consensus (VBSBC) algorithms demonstrates that as the number of attacks increases, detection rates must correspondingly improve to maintain network security [37]. DoS attacks in autonomous vehicle contexts could target critical systems such as emergency braking protocols, route recalculation services, or communication with emergency response services, effectively disabling autonomous vehicles during critical operational moments when manual intervention may not be possible [12].

3.4 *Quantum Computing Threats and Cryptographic Vulnerabilities*

The emergence of quantum computing capabilities represents a fundamental threat to the cryptographic foundations underlying smart contract security in autonomous vehicles [38]. Shor's quantum factoring algorithm can efficiently solve the Elliptic Curve Discrete Logarithm Problem (ECDLP) and Integer Factorization Problem (IFP) that form the basis of current digital signature schemes, including ECDSA, RSA, and EdDSA [39]. In autonomous vehicle applications, quantum adversaries could potentially derive private keys from public keys, forge signatures for vehicle authentication or smart contract execution, manipulate consensus mechanisms in Proof of Stake systems, or compromise smart contracts that depend on signature-based permissions. The "harvest now, exploit later" scenario presents particularly concerning implications for autonomous vehicle deployments. Malicious actors could record blockchain transactions and smart contract interactions from current autonomous vehicle operations, then decrypt or forge signatures once quantum computing capabilities advance sufficiently to break existing cryptographic protections [40]. This temporal vulnerability means that sensitive autonomous vehicle data and financial transactions recorded today could become compromised years later, potentially exposing historical location patterns, payment information, and operational data.

3.4.1 Side Channel Attacks and Physical Vulnerabilities

Autonomous vehicles face unique side channel attack vulnerabilities due to their reliance on shared hardware platforms for critical computational tasks. Cache-based side channel attacks can potentially compromise location privacy by exploiting shared hardware resources between autonomous driving software and attack programs [7]. Research demonstrates that adaptive behaviors in perception and control algorithms may introduce novel side channel vulnerabilities that reveal physical properties of vehicles or their environmental context [41]. Statistical learning models can infer states from Adaptive Monte-Carlo Localization (AMCL) algorithms through cache access pattern analysis, potentially allowing attackers to predict vehicle routes or locations from algorithmic state traces [41].

3.4.2 Hardware Performance Counter-Based Attacks

Hardware performance counter-based attacks represent another sophisticated threat vector where attackers leverage microarchitectural information to fingerprint autonomous vehicle operations [7]. These attacks can potentially identify specific routes, detect passenger patterns, or even compromise deep learning model inputs used in autonomous vehicle decision-making systems [30]. The correlation between sensitive information and side channel observations highlights critical vulnerabilities in autonomous vehicle computational architectures that traditional security measures may not address.

3.5 Detection and Mitigation Challenges

Current vulnerability detection methodologies demonstrate significant limitations when applied to autonomous vehicle smart contract environments. Traditional tools, including Slither, Mythril, MythX, and Oyente, frequently produce false positives and operate within restricted environmental constraints that may not accurately reflect the dynamic operational conditions of autonomous vehicles [5]. The VulnScan GPT framework represents recent advances in automated vulnerability detection, combining Abstract Syntax Tree analysis with vector database storage and GPT-based analysis to improve detection accuracy while reducing operational costs [42]. However, these tools still face challenges in handling function segmentation complexities, signature similarity disambiguation, and the uncertainty inherent in autonomous vehicle decision-making systems. Formal verification approaches offer promising solutions for ensuring autonomous vehicle smart contract security, but face scalability limitations when applied to complex, real-time distributed systems [15]. The integration of Runtime Verification (RV) techniques for monitoring smart contract execution in autonomous vehicle environments requires adaptation to handle the uncertainty and non-deterministic behavior characteristic of AI-based vehicle control systems [43]. The heterogeneous composition of autonomous vehicle systems, combining white-box open-source components with black-box proprietary systems, presents additional verification challenges that require specialized assume-guarantee verification methodologies [7].

3.6 *Consequences and Risk Assessment*

The consequences of smart contract vulnerabilities in autonomous vehicle environments extend far beyond traditional blockchain security concerns. While conventional smart contract failures primarily result in financial losses, autonomous vehicle smart contract vulnerabilities can directly threaten human life and public safety [3]. The safety-critical nature of autonomous vehicle operations demands unprecedented levels of security assurance that current smart contract development practices and security frameworks may not adequately provide. The immutable nature of blockchain technology, while providing security benefits, simultaneously means that vulnerabilities embedded within deployed smart contracts become permanent fixtures that cannot be easily rectified without complex upgrade procedures, creating long-term security risks for autonomous vehicle fleets [11]. High-speed vehicle mobility, rapidly changing network topologies, and real-time decision-making requirements create unique attack surfaces that require specialized security approaches specifically designed for autonomous vehicle environments [13]. As autonomous vehicle technology continues to evolve and quantum computing capabilities advance, the security challenges facing smart contract implementations in AIV systems will likely become increasingly sophisticated, demanding proactive and adaptive security strategies to ensure safe and secure autonomous transportation systems.

4 Quantum Intelligence for Addressing Smart Contract Security Problems

Security assurance of smart contracts is important to ensure the safety of AV, especially in level 5, where manual control will no longer be necessary for AV operation; therefore, in AV, there will be minimal options for manual control. Therefore, security breaches in smart contract in AV operation can causes serious problems in the overall vehicular system. To detect vulnerability in smart contract, several commercial tools are available for example, Slither [44], Securify [45], Oyente [46], and others. Many others research based methods also available for vulnarable detection [47, 48], and others. These tools mainly perform static analysis, fuzz testing, symbolic execution, and interactive debugging for vulnerability detection. However, these testing tools are not always produce accurate vulnerability detection, often produce false positives, limited to specific environments, and other issues.

Therefore, deep learning (DL)-based smart contract security detection method is getting popular day by day due to its higher accuracy and completeness [49]. Different types of DL approaches are applied for vulnerability detection for example Long Short Term Memory (LSTM), bidirectional GRU (Bi-GRU), transfer learning, and others approaches. However, introducing quantum computing to secure smart contract from different kinds of attack, could be a game-changer. Quantum computing is already showing its supremacy over classical computers in many fields despite the extreme hardware limitations. Quantum computing is a new way of computing, different foundamentally from classical computers. Unlike classical bits that are either 0 or 1, a qubit can exist in a superposition of 0 and 1 and can become entangled with other qubits. Quantum algorithms exploit superposition, entanglement, and interference to perform certain computations more efficiently than classical algorithms. Quantum machine learning already using in differnt field of AU functionalities, network intrusion detection, vehicle routing path planning and optimization, traffic prediction and decision making and others critical tasks.

In addition to detecting logical, structural, and syntax vulnerabilities, encryption plays a crucial role in smart contract security. Current systems rely on classical encryption techniques such as public key cryptography (RSA, Elliptic Curve Cryptography), hashing algorithms, and digital signatures. However, these techniques may become vulnerable to powerful quantum computers. Shor's algorithm threatens RSA and elliptic curve cryptography by enabling efficient key recovery, while Grover's algorithm reduces the effective security margin of hash functions, which motivates longer digests and conservative parameters. In response, many blockchain platforms are actively developing and upgrading toward quantum resistance. For instance, Ethereum 2.0 is expected to be quantum-proof, Pure Chain is integrating quantum encryption, and other platforms are pursuing similar advancements. Quantum-enabled algorithms such as Quantum Key Distribution (QKD) and quantum digital signatures are potential candidates for future encryption methods. Nevertheless, the key generation time for these algorithms is relatively high, making them unsuitable for commercial and high-throughput applications like autonomous intelligent vehicle (AIV) security. Consequently, post-quantum encryption and quantum-enabled encryption for smart contracts have emerged as active areas of research.

5 Quantum-Based Distributed Communication in V2X, V2I, V2E Architecture

Connectivity is an important factor for AV. AV needs to communicate seemlessly with another AV, infrastructure, and control network for smooth operation. Depending on the type of operation, vehicular network includes vehicle-to-vehicle (V2V), vehicle-to-network (V2N), vehicle-to-event (V2E), vehicle-to-Infrastructure

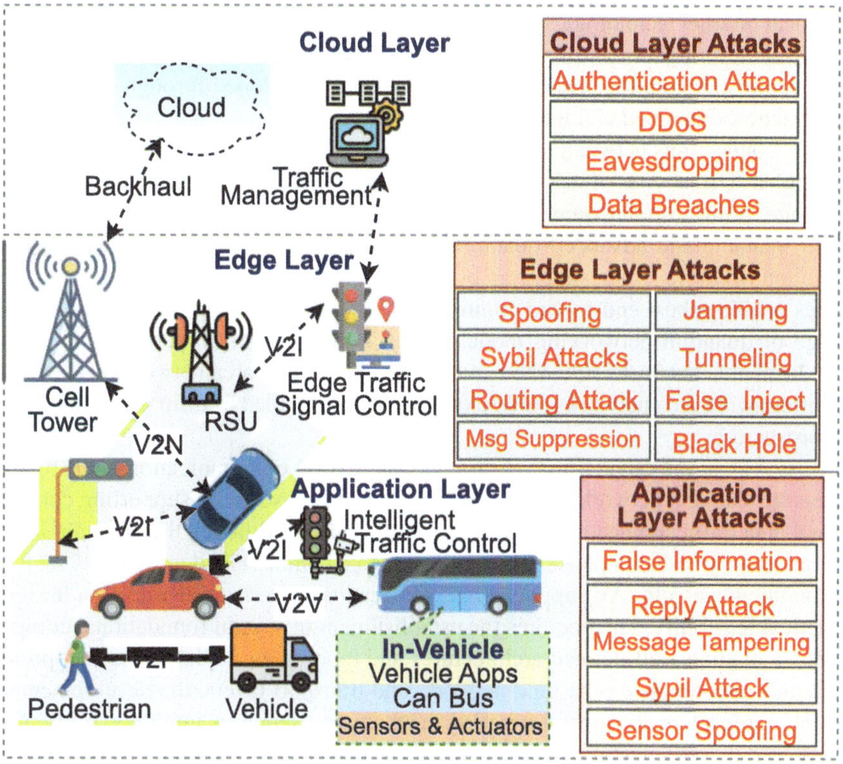

Fig. 1 V2X network illustrating potential attacks at each layer

(V2I), vehicle-to-pedestrian (V2P), and the combination called vehicle-to-everything (V2X). Figure 1 shows different types of attacks in different types of vehicualar communication networks. There is possibility of different types of attacks in each type of vehicular network. In the case of AV operations, especially level 3 and beyond, any security breach could have a serious effect.

For AV operation, vehicle communication needs to maintain strict security criteria within network topology, among other factors like ultra-reliability, low latency, efficient data handling, and others. There is a huge possibility is waiting of quantum computation integrated V2X communication. The application of quantum computing in V2X or V2E could be quantum internet, quantum qncryption enhanced vehicular network, or quantum optimization enabled network traffic management and optimization.

The quantum internet is built on the principles of quantum physics, such as teleportation, superposition, and entanglement. Due to the no-cloning property of quantum bits (qubits), any eavesdropping attempt on the quantum internet can be detected. In addition to this, the quantum internet offers several unique properties, while current classical internet technologies and hardware are approaching their performance limits. To achieve higher bandwidth and enhanced security, the quantum internet could be a promising solution. Quantum internet can be developed through two distinct approaches: entangled and unentangled architectures. In unentangled quantum networks, connections between nodes are established using quantum states that are not entangled. These types of networks are particularly suitable for implementing short-distance, point-to-point quantum key distribution (QKD) protocols. On the other hand, scalable and advanced quantum internet systems rely heavily on entangled quantum networks, often supported by quantum repeaters. Long-distance entangled states enable robust end-to-end communication and hold great promise for a wide range of quantum networking models. Such networks can be deployed as quantum local area networks (QLANs), quantum metropolitan area networks (QMANs), quantum wide area networks (QWANs), and even extended to form global quantum infrastructures.

It is widely anticipated that sixth-generation (6G) communications will tightly integrate classical networking with quantum capabilitiesthereby supporting conventional data traffic alongside quantum-enabled services. Although the quantum internet in an early research phase. Prior to large-scale commercial deployments such as autonomous vehicle (AV) applications, the quantum internet must reach a level of practical feasibility. This requires the establishment of several foundational components, including well-defined architectures and topologies, a quantum-aware protocol stack encompassing the link, network, and transport layers, the development of quantum repeaters and error-correction mechanisms, seamless interoperability with classical networks, and, most critically, the formulation of internationally recognized standards.

6 Challenges of Quantum and Post-Quantum Intelligence Integration in AIV Security

Implementing quantum-enabled intelligence in Autonomous Intelligent Vehicles (AIVs) requires overcoming several challenges, the most significant is hardware limitations. Although quantum computing is not a new concept, quantum computers have not yet reached their first generation. Qubits, the fundamental units of quantum computation, are highly vulnerable, and constructing memory using qubits remains a difficult task. Another major hardware challenge is that quantum computers must operate at extremely low temperatures, near absolute zero, in order to preserve the quantum state of qubits as dictated by quantum physics. However, an area of active research is the development of quantum hardware that operates at room temperature. This field is steadily progressing, with some studies claiming successful development of room-temperature quantum chips, although these technologies are not yet commercially available. Quantum hardware is advancing rapidly, pushing the field from today's noisy intermediate-scale quantum (NISQ) era toward future fault-tolerant quantum computing (FTQC), though large-scale FTQC has not yet been achieved. However, current security standards for autonomous vehicles, such as ISO 26262 and ISO/PAS 21448, do not incorporate quantum computing. Instead, they continue to focus on traditional classical approaches. Therefore, the security standards must be updated to incorporate quantum-enabled encryption and post-quantum encryption in Autonomous Intelligent Vehicle (AIV) operations. Additionally, it is essential to integrate quantum-enabled security solutions within the software infrastructure, which will require substantial upgrades to accommodate quantum intelligence. As a result, the development and deployment of these advanced technologies will necessitate the creation of a skilled workforce, with experts specialized in quantum intelligence, to ensure the effective implementation and maintenance of these systems.

7 Case Study: Quantum-Based Smart Contract Implementation in AIVs

As stated beginning in this chapter, to ensure the safe operation of AIVs with the integration of smart contract and blockchain, it is critical to ensure that the smart contract itself is secure. This section explores a novel vulnerability detection mechanism using quantum transformer learning in smart contracts. Transformer learning, a significant approach within machine learning, processes input data and transforms it into the desired output. Its ability to manage sequences and contextual information makes it a foundational method for generative artificial intelligence (Gen AI), which is the ground model capable of generating new content based on learned data patterns.

Considering blockchain-integrated platforms, the security of smart contracts is important, as any vulnerability could compromise the integrity of the entire system. Quantum transformer learning represents a cutting-edge methodology that leverages quantum computing principles to enhance the detection of such vulnerabilities. The hybrid approach introduced in this work combines quantum mechanics with transformer learning, providing a potent tool to identify potential weaknesses in smart contract code before they can be exploited. Sections 3 and 4 of this chapter analyze deeper into the details of machine learning-based vulnerability detection mechanisms in smart contracts. Here, we introduce the hybrid quantum transformer learning model, which integrates quantum algorithms with traditional transformer architectures. This method is designed to detect vulnerabilities by utilizing the computational power of quantum processing. The system is a case study of evaluating potential threats more effectively, offering a higher level of security for blockchain applications, particularly in the context of AV operations where real-time, robust security measures are critical.

The hybrid quantum transformer learning-based smart contract vulnerability detection model utilizes a publicly available dataset [50]. The data in this dataset were collected from real-world smart contracts deployed on the Ethereum blockchain

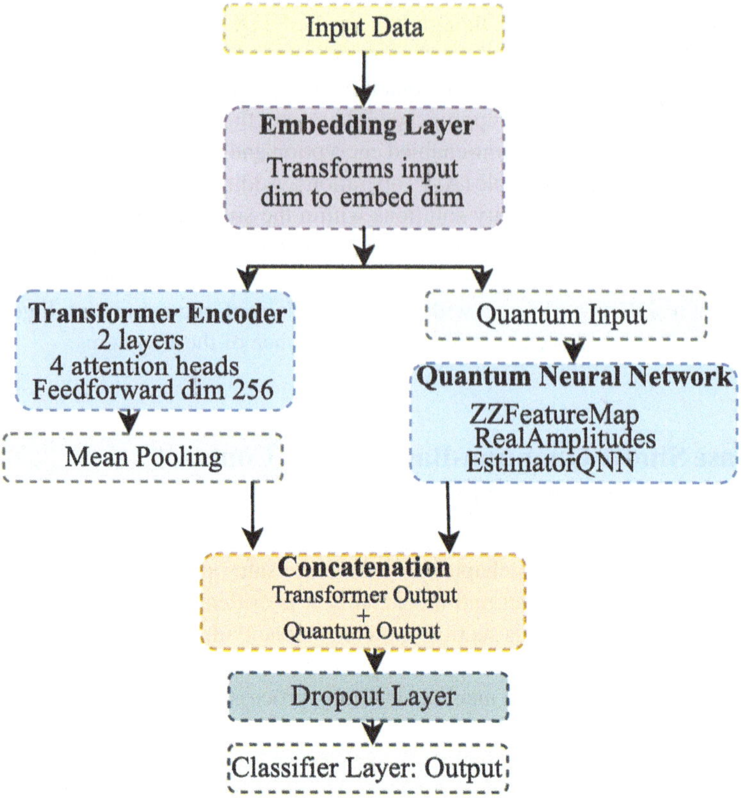

Fig. 2 Diagram of the hybrid quantum transformer learning model for smart contract vulnerability detection

network and VNT Chain platforms. The vulnerabilities identified within the dataset include various types, such as Timestamp Dependence (TD), Reentrancy, Integer Overflow/Underflow, Dangerous Delegatecall, among others [51]. A tokenization and vectorization approach is applied to process textual data for machine learning tasks. The TfidfVectorizer is employed to transform the raw text data into a numerical format by computing the Term Frequency-Inverse Document Frequency (TF-IDF) of the words in the text.

Figure 2 illustrates a hybrid quantum transformer model designed for smart contract vulnerability detection. It begins with the input data, which is passed through an embedding layer that transforms the input dimension to the embedding dimension. The embedded data then enters the Transformer Encoder with two layers, utilizing four attention heads and a feedforward dimension. Following this, the data is processed by a Quantum Neural Network (QNN) consisting of a ZZFeatureMap for feature encoding and a RealAmplitudes ansatz. The quantum network processes the

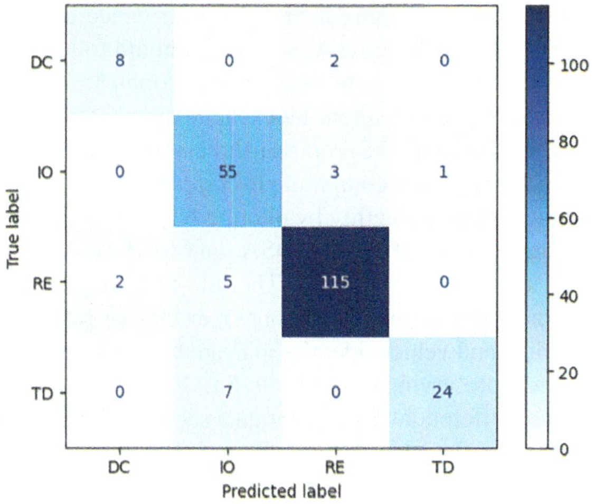

Fig. 3 Confusion matrix of vulnerability detection of the hybrid quantum transformer learning model

data and produces a quantum output, which is then concatenated with the classical output from the Transformer Encoder. This combined representation is passed through a dropout layer for regularization and finally fed into a classifier layer that outputs the model's predictions.

Figure 3 shows the vulnerability detection output of the hybrid quantum transformer learning model in the confusion matrix. The results indicate a strong predictive capability of the four classes. The model shows accuracy above 91% in 15 epoch with a precision 0.9051, and F1 Score 0.9059. Compared to the results in [52], which achieved an accuracy of 89.57%, a precision of 84.5%, and an F1 score of 83.55, our proposed method demonstrates outperformance. However, the deployment of the hybrid quantum transformer learning-based smart contract vulnerability detection system have several challenges. It requires access to quantum processors or simulators, which can be costly and limited in availability. Integrating quantum algorithms with classical machine learning may also lead to compatibility issues, demanding specialized expertise. The system's reliance on publicly available datasets raises concerns about data quality and generalization to new vulnerabilities. Real-time processing, essential for blockchain applications like autonomous vehicles, may be hindered by the system's computational demands.

8 Conclusion

This chapter explored how blockchain and smart contracts are reshaping cybersecurity in autonomous vehicles. While smart contracts offer new capabilities, they also introduce unique security challenges. Quantum computing, along with other classical advancements, adds further complexity to these challenges. Therefore, transitioning AIVs toward quantum resilience is essential to meet both current and future security requirements. However, since quantum computing remains in the NISQ stage, enhancing AIV security should follow a gradual, step-by-step approach aligned with the pace of quantum development. The transition of the AIV ecosystem onto a quantum-resilient blockchain foundation demands a disciplined, standards-aligned roadmap. We propose a three-stage transition to mitigate the existential threat posed by quantum computing to current cryptographic systems: a near-term focus on achieving crypto-agility by piloting NIST Post-Quantum Cryptography (PQC) standards like ML-KEM, ML-DSA, and SLH-DSA for key establishment, signatures, and long-lived attestations. The mid-term requires integrating PQC as the default for critical functions, including Over-the-Air (OTA) updates, software supply chain signing, and vehicle identity management, while simultaneously adopting advanced privacy-preserving tools like zk-STARKs for compliance and homomorphic encryption or differential privacy for data services. Finally, the long-term phase culminates in the complete deprecation of classical public-key cryptography, migration to PQC-native roots, and the finalization of incident-response playbooks. Architecturally, this shift is embodied by a hybrid blockchain–quantum stack featuring permissioned BFT consensus secured by PQC, STARK-based verifiability, and encrypted off-chain computation protected by PQC Key Encapsulation Mechanisms (KEMs). Operationalizing this transition necessitates the establishment of regulatory sandboxes (aligned with UN R155 and ISO/SAE 21434) and a pre-competitive AIV-PQC consortium to develop shared, standardized assurance artifacts and reference profiles, thereby proactively balancing security, verifiability, privacy, and commercial practicality in the face of quantum-era risk.

Supplementary information

If your article has accompanying supplementary file/s please state so here.

Authors reporting data from electrophoretic gels and blots should supply the full unprocessed scans for key as part of their Supplementary information. This may be requested by the editorial team/s if it is missing.

Please refer to Journal-level guidance for any specific requirements.

Declarations Conflict of interest On behalf of all authors, the corresponding author states that there is no conflict of interest.

References

1. Committee, J.: Taxonomy and definitions for terms related to driving automation systems for on-road motor vehicles. SAE Int. **J3016**(202104) (2021). https://doi.org/10.4271/J3016_202104
2. Gros, C., Kester, L., Martens, M., et al.: Addressing ethical challenges in automated vehicles: bridging the gap with hybrid ai and augmented utilitarianism. AI Ethics **5**, 2757–2770 (2025). https://doi.org/10.1007/s43681-024-00592-6

3. Kontos, C., Panagiotakopoulos, T., Kameas, A.: Applications of blockchain and smart contracts to address challenges of cooperative, connected, and automated mobility. Sensors (Basel, Switzerland) **24**(19), 6273 (2024)
4. Ajakwe, S.O., Okoloegbo, C.A., Lee, J.M., Kim, D.S.: Machine learning models for drone security: cognitive versus cyber intelligence for safety operations. In: Machine Learning for Drone-Enabled IoT Networks: Opportunities, Developments, and Trends, pp. 121–139. Springer, Berlin (2025)
5. Wang, X., Zhu, H., Ning, Z., Guo, L., Zhang, Y.: Blockchain intelligence for internet of vehicles: challenges and solutions. IEEE Commun. Surv. Tutor. **25**(4), 2325–2355 (2023)
6. Ajakwe, S.O., Igboanusi, I.S., Lee, J.-M., Kim, D.-S.: iBANDA: a blockchain-assisted defense system for authentication in drone-based logistics. Drones **9**(8), 590 (2025)
7. Tyagi, A.K., Sreenath, N.: Future intelligent vehicles: open issues, critical challenges, and research opportunities. In: Intelligent Transportation Systems: Theory and Practice, pp. 327–358. Springer, Berlin (2022)
8. Ajakwe, S.O., Kim, D.-S., Lee, J.-M.: Drone transportation system: systematic review of security dynamics for smart mobility. IEEE Internet Things J. **10**(16), 14462–14482 (2023)
9. Lin, H.-Y.: A secure transaction and smart contract architecture for internet of vehicles. Sci. Prog. **107**(4), 00368504241296039 (2024)
10. Lara-Nino, C.A., Diaz-Perez, A., Morales-Sandoval, M.: Elliptic curve lightweight cryptography: a survey. IEEE Access **6**, 72514–72550 (2018)
11. Singh, P.K., Singh, R., Nandi, S.K., Ghafoor, K.Z., Rawat, D.B., Nandi, S.: Blockchain-based adaptive trust management in internet of vehicles using smart contract. IEEE Trans. Intell. Transp. Syst. **22**(6), 3616–3630 (2020)
12. Jain, S., Ahuja, N.J., Srikanth, P., Bhadane, K.V., Nagaiah, B., Kumar, A., Konstantinou, C.: Blockchain and autonomous vehicles: recent advances and future directions. IEEE Access **9**, 130264–130328 (2021)
13. Wu, H., Yue, C., Zhang, L., Li, Y., Imran, M.A.: When distributed consensus meets wireless connected autonomous systems: a review and a DAG-based approach. IEEE Netw. (2024)
14. Ajakwe, S.O., Olabisi, K.L., Kim, D.-S.: Multihop intruder node detection scheme (MINDS) for secured drones' FANET communication. IET Intel. Transp. Syst. **19**(1), 70080 (2025)
15. Javaid, U., Sikdar, B.: A secure and scalable framework for blockchain based edge computation offloading in social internet of vehicles. IEEE Trans. Veh. Technol. **70**(5), 4022–4036 (2021)
16. Distefano, S., Di Giacomo, A., Mazzara, M.: Trustworthiness for transportation ecosystems: the blockchain vehicle information system. IEEE Trans. Intell. Transp. Syst. **22**(4), 2013–2022 (2021)
17. Surapaneni, P., Bojjagani, S., Khan, M.K.: Dynamic-trust: Blockchain-enhanced trust for secure vehicle transitions in intelligent transport systems. IEEE Trans. Intell. Transp. Syst. (2025)
18. Aman, M.N., Javaid, U., Sikdar, B.: A privacy-preserving and scalable authentication protocol for the internet of vehicles. IEEE Internet Things J. **8**(2), 1123–1139 (2020)
19. Iqbal, M., Suhail, S., Matulevičius, R., Shah, F.A., Malik, S.U.R., McLaughlin, K.: IoV-TwinChain: predictive maintenance of vehicles in internet of vehicles through digital twin and blockchain. Internet Things **30**, 101514 (2025)
20. Sahal, R., Alsamhi, S.H., Brown, K.N., O'shea, D., McCarthy, C., Guizani, M.: Blockchain-empowered digital twins collaboration: smart transportation use case. Machines **9**(9), 193 (2021)
21. Kazmi, S.A., Dang, T.N., Yaqoob, I., Manzoor, A., Hussain, R., Khan, A., Hong, C.S., Salah, K.: A novel contract theory-based incentive mechanism for cooperative task-offloading in electrical vehicular networks. IEEE Trans. Intell. Transp. Syst. **23**(7), 8380–8395 (2021)
22. Ly, R., Shojaei, A.: Decentralized autonomous organization in built environments: applications, potential and limitations. Inf. Syst. e-Bus. Manag. 1–46 (2025)
23. Vitiello, V., Benazzi, A., Trucillo, P.: Smart card-based vehicle ignition systems: security, regulatory compliance, drug and impairment detection, through advanced materials and authentication technologies. Processes **13**(3), 911 (2025)

24. Bader, L., Bürger, J.C., Matzutt, R., Wehrle, K.: Smart contract-based car insurance policies. In: 2018 IEEE Globecom Workshops (GC Wkshps), pp. 1–7. IEEE (2018)
25. Duan, Y., Zhao, X., Pan, Y., Li, S., Li, M., Xu, F., Zhang, M.: Towards automated safety vetting of smart contracts in decentralized applications. In: Proceedings of the 2022 ACM SIGSAC Conference on Computer and Communications Security, pp. 921–935 (2022)
26. Huang, K., Parisi, C., Tan, L.J., Ma, W., Zhang, Z.W.: Web3 Applications Security and New Security Landscape. Springer, Berlin (2024)
27. Giannaros, A., Karras, A., Theodorakopoulos, L., Karras, C., Kranias, P., Schizas, N., Kalogeratos, G., Tsolis, D.: Autonomous vehicles: sophisticated attacks, safety issues, challenges, open topics, blockchain, and future directions. J. Cybersecur. Priv. **3**(3), 493–543 (2023)
28. Kanj, S., Garcia, P., Rosés, O., Pegueroles, J.: A review of tactics, techniques, and procedures (TTPs) of MITRE framework for business email compromise (BEC) attacks. IEEE access (2025)
29. Soofiyan, S., Karami, A.: Ethereum smart contracts: a hierarchical analysis of vulnerability challenges and mitigation strategies. Clust. Comput. **28**(8), 543 (2025)
30. Kidmose, B.: A review of smart vehicles in smart cities: dangers, impacts, and the threat landscape. Veh. Commun. 100871 (2024)
31. Choy, W., Teng, P.: When smart contracts are outsmarted: the parity wallet 'Freeze' and software liability in the internet of value. Natl. Law Rev. **7**(356) (2017)
32. Peng, L., Feng, W., Yan, Z., Li, Y., Zhou, X., Shimizu, S.: Privacy preservation in permissionless blockchain: a survey. Digit. Commun. Netw. **7**(3), 295–307 (2021). https://doi.org/10.1016/j.dcan.2020.05.008
33. Alabdulatif, A., Alharbi, M., Mchergui, A., Moulahi, T.: Mitigating blackhole and greyhole routing attacks in vehicular ad hoc networks using blockchain based smart contracts. CMES-Comput. Model. Eng. Sci. **140**(2) (2024)
34. Maleknasab Ardakani, M., Tabarzad, M.A., Shayegan, M.A.: Detecting sybil attacks in vehicular ad hoc networks using fuzzy logic and arithmetic optimization algorithm. J. Supercomput. **78**(14), 16303–16335 (2022)
35. Hataba, M., Sherif, A., Mahmoud, M., Abdallah, M., Alasmary, W.: Security and privacy issues in autonomous vehicles: a layer-based survey. IEEE Open J. Commun. Soc. **3**, 811–829 (2022)
36. Platt, M., McBurney, P.: Sybil attacks on identity-augmented proof-of-stake. Comput. Netw. **199**, 108424 (2021)
37. Shen, X., Ma, R.: A blockchain solution for the internet of vehicles with better filtering and adaptive capabilities. Sensors **25**(4), 1030 (2025)
38. Ajakwe, S.O., Kim, D.-S.: EQAI: explainable quantum-empowered antispoofing intelligence for trustworthy connected autonomous vehicles communication. IEEE Internet Things J. 1 (2025) https://doi.org/10.1109/JIOT.2025.3638378
39. Roetteler, M., Naehrig, M., Svore, K.M., Lauter, K.: Quantum resource estimates for computing elliptic curve discrete logarithms. In: International Conference on the Theory and Application of Cryptology and Information Security, pp. 241–270. Springer, Berlin (2017)
40. Pradhan, T., Patil, P.: Quantum cryptography for secure autonomous vehicle networks: a review. In: 2024 IEEE International Students' Conference on Electrical, Electronics and Computer Science (SCEECS), pp. 1–10. IEEE (2024)
41. Pendleton, S.D., Andersen, H., Du, X., Shen, X., Meghjani, M., Eng, Y.H., Rus, D., Ang, M.H.: Perception, planning, control, and coordination for autonomous vehicles. Machines **5**(1), 6 (2017)
42. Wang, L., Zhao, Y., Wang, K., Zhang, Z., Zhang, W.: VulnScan GPT: a new framework for smart contract vulnerability detection combining vector database and GPT model. J. Surveill., Secur. Saf. **5**(4), 258–272 (2024)
43. Caldas, R., García, J.A.P., Schiopu, M., Pelliccione, P., Rodrigues, G., Berger, T.: Runtime verification and field-based testing for ROS-based robotic systems. IEEE Trans. Softw. Eng. (2024)
44. Feist, J., Grieco, G., Groce, A.: Slither: a static analysis framework for smart contracts. In: 2019 IEEE/ACM 2nd International Workshop on Emerging Trends in Software Engineering for Blockchain (WETSEB), pp. 8–15 (2019). https://doi.org/10.1109/WETSEB.2019.00008

45. Tsankov, P., Dan, A., Drachsler-Cohen, D., Gervais, A., Buenzli, F., Vechev, M.: Securify: practical security analysis of smart contracts. In: Proceedings of the 2018 ACM SIGSAC Conference on Computer and Communications Security, pp. 67–82 (2018)
46. Luu, L., Chu, D.-H., Olickel, H., Saxena, P., Hobor, A.: Making smart contracts smarter. In: Proceedings of the 2016 ACM SIGSAC Conference on Computer and Communications Security, pp. 254–269 (2016)
47. Chen, J., Xia, X., Lo, D., Grundy, J., Luo, X., Chen, T.: Defining smart contract defects on ethereum. IEEE Trans. Softw. Eng. **48**(1), 327–345 (2020)
48. Jin, H., Wang, Z., Wen, M., Dai, W., Zhu, Y., Zou, D.: AROC: an automatic repair framework for on-chain smart contracts. IEEE Trans. Softw. Eng. **48**(11), 4611–4629 (2021)
49. Chu, H., Zhang, P., Dong, H., Xiao, Y., Ji, S.: Deepfusion: smart contract vulnerability detection via deep learning and data fusion. IEEE Trans. Reliab. **74**(3), 3544–3558 (2025). https://doi.org/10.1109/TR.2024.3480010
50. Zhuang, Y., Liu, Z., Qian, P., Liu, Q., Wang, X., He, Q.: Smart contract vulnerability detection using graph neural networks. In: Proceedings of the Twenty-ninth International Conference on International Joint Conferences on Artificial Intelligence, pp. 3283–3290 (2021)
51. Messi-Q: Smart Contract Dataset (2021). https://github.com/Messi-Q/Smart-Contract-Dataset. Accessed 30 Sept 2025
52. Yang, Y., Xu, H., Yao, S.: Enhancing smart contract vulnerability detection with hybrid pre-trained models on source code and byte code. In: 2024 6th International Conference on Frontier Technologies of Information and Computer (ICFTIC), pp. 122–126. IEEE (2024)

Blockchain Trilemma in Autonomous Intelligent Vehicle

Chimeremma Sandra Amadi and Taesoo Jun

Abstract Autonomous Intelligent Vehicles (AIVs) require secure, low-latency, and scalable communication frameworks to operate safely in dense and dynamic environments. Blockchain, with its decentralized and tamper-resistant ledger, offers a promising foundation for enhancing trust in vehicle to vehicle (V2V) and Vehicle to Infrastructure (V2I) systems. Yet, the blockchain trilemma, balancing security, speed, and scalability, poses significant challenges in safety-critical mobility contexts. This chapter examines how the trilemma manifests in AIV ecosystems, where high throughput event logging, real-time consensus, and hardware constraints intensify trade-offs. We present innovation driven approaches, including hybrid consensus protocols, edge-enabled blockchain layers, lightweight cryptographic primitives, and AI-based modules for anomaly detection and validator prediction. Two case studies of urban intersections and highway platooning illustrate the practical tensions between latency, security, and scalability, and demonstrate how tiered blockchain architectures can mitigate these risks.

Keywords Autonomous intelligent vehicles · Blockchain trilemma · V2V · V2I · Security · Scalability · Speed

1 Introduction

The rapid evolution of AIVs is reshaping the future of mobility by integrating artificial intelligence, advanced sensing, and high-speed communication technologies. Unlike conventional vehicles, AIVs are designed to operate in V2V and V2I networks where safety, efficiency, and reliability depend on constant real-time data

C. S. Amadi · T. Jun
Department of IT Convergence Engineering, Kumoh National Institute of Technology, Gumi, South Korea
e-mail: taesoo.jun@kumoh.ac.kr

C. S. Amadi (✉)
Department of Information Technology, Federal University of Technology, Owerri, Nigeria
e-mail: chimesandra@yahoo.com

S. O. Ajakwe and D.-S. Kim (eds.), *Blockchain for Security of Autonomous Intelligent Vehicles*, Studies in Big Data 194,
https://doi.org/10.1007/978-3-032-20851-4_8

exchange. In dense urban traffic, highway platooning, or mixed traffic environments, vehicles must process vast amounts of sensor data and make decisions within milliseconds. Consequently, AIV ecosystems demand communication frameworks that are simultaneously secure, to protect against malicious interference; fast, to guarantee real-time responsiveness; and scalable, to support the growing number of connected vehicles entering smart transport infrastructures [1–4]. Blockchain has emerged as a promising enabler of trust and security in AIV networks. Its decentralized ledger ensures data integrity by preventing unauthorized tampering, while cryptographic primitives guarantee authentication and identity management across distributed nodes [5]. Unlike traditional centralized systems, blockchain eliminates single points of failure, making vehicular communication more resilient to disruption or malicious attacks. Furthermore, transparency and immutability in blockchain fosters trust among heterogeneous stakeholders, including manufacturers, service providers, and vehicles themselves. As AIV adoption expands, trust frameworks that operate across organizational and jurisdictional boundaries become indispensable. Blockchain provides precisely this shared, verifiable record of transactions and events [6]. Recent research further highlights the advantages of AI blockchain integration for vehicular networks [7]. Artificial intelligence algorithms enhance blockchain functionality by enabling anomaly detection in vehicular data, predictive validation of network behavior, and adaptive consensus optimization [8]. This integration addresses challenges such as detecting fraudulent messages, predicting node misbehavior, and optimizing communication overhead. For example, AI-driven anomaly detection can reduce false positives in vehicular threat detection, while blockchain guarantees the immutability of alerts, thereby improving both security and reliability in AIV decision-making [4, 9, 10].

Despite these advantages, blockchain adoption in AIV systems faces the enduring challenge of the blockchain trilemma, the difficulty of simultaneously achieving security, scalability, and speed (throughput/latency) in a single system [11–15]. Originally conceptualized as a trade-off among security, scalability, and decentralization, the trilemma manifests differently in vehicular contexts. For AIVs, decentralization remains relevant but is overshadowed by the immediate need to reconcile security, speed, and scalability, all of which directly affect system safety and operational performance. This tension is evident in existing blockchain platforms. For instance, Bitcoin prioritizes security and decentralization through proof-of-work consensus but sacrifices transaction throughput, operating at fewer than ten transactions per second [2, 10]. Such low throughput is unacceptable for vehicular environments that require near-instantaneous validation of thousands of data exchanges per second. Attempts to enhance scalability through mechanisms such as sharding, state channels, and side chains have shown improvements in throughput but often weaken consensus robustness or security guarantees [16, 17]. Importantly, theoretical research suggests that these trade-offs are not merely engineering compromises but reflect fundamental computational limits, with proofs indicating the inevitability of such tensions [14].

For AIV applications, the implications of this trilemma are profound. Security cannot be compromised, as attacks on vehicular communication could lead to catastrophic outcomes such as collisions or system-wide failures. Speed is equally critical, since high-latency consensus protocols cannot support the real-time decision-making required for collision avoidance or cooperative maneuvers. Finally, scalability must be ensured to accommodate the rapid growth of connected vehicles in dense urban environments and highway networks. Addressing all three dimensions in tandem is thus one of the central challenges for blockchain-enabled AIV systems [3, 5]. Therefore, it becomes essential to critically examine how these tensions manifest in vehicular ecosystems and explore targeted design strategies for resolution. This chapter critically engages with these issues and provides a structured examination of how the blockchain trilemma constrains AIV adoption, while also outlining innovative approaches for resolution. Specifically, the chapter will contribute in the following ways:

- Dissect the blockchain trilemma in the context of AIVs, analyzing how the competing demands of security, speed, and scalability manifest in vehicular networks and how they hinder the adoption of blockchain-based solutions.
- Present innovative design techniques for securing blockchain-enabled AIV systems, including AI-driven enhancements and architectural approaches that strengthen vehicular network resilience while preserving performance.
- Propose scalable frameworks and connectivity strategies that reconcile the trade-offs between security, speed, and scalability, offering practical pathways for deploying blockchain in large-scale AIV ecosystems.

2 Understanding the Blockchain Trilemma in AIV

The blockchain trilemma, coined by Vitalik Buterin, highlights the inherent difficulty of simultaneously achieving security, decentralization, and scalability in distributed ledger systems. It suggests that optimizing two of these properties typically comes at the expense of the third, creating a fundamental design trade-off in blockchain networks [18]. For example, systems emphasizing strong decentralization and security often struggle with throughput limitations, while those prioritizing scalability may compromise either security or decentralization. Originally conceived to support peer-to-peer cryptocurrency transactions, blockchain was designed as a decentralized ledger that ensured trust and transparency through encryption-based security. Over time, this paradigm has been extended beyond financial applications to intelligent transportation systems, where each vehicle can function as a node within a decentralized network [19]. In such a setting, blockchain offers clear benefits including enhanced cybersecurity against vehicle hacking, adaptive resilience as the network grows, and trustworthy information exchange without reliance on centralized authorities or servers.

Table 1 Comparison between classical blockchain trilemma and AIV-adopted trilemma

Dimension	Classical blockchain trilemma	AIV-adopted trilemma
Security	Protection against double-spending, data tampering, and consensus manipulation	Cryptographic integrity of vehicular data, resistance to replay and Sybil attacks, secure V2V and V2I communication.
Decentralization/Speed	Decentralization is essential for trustless consensus but often slows down throughput and increases latency	Speed is prioritized over decentralization, requiring real-time or near-real-time consensus for collision avoidance and cooperative maneuvers
Scalability	Ability to process a high volume of transactions without reducing security or decentralization	Ability to support thousands of AIVs in dense traffic with continuous high-frequency event logging while preserving system responsiveness

When transposed into the context of AIV, the trilemma requires reframing. While decentralization continues to play a role in avoiding single points of failure and ensuring independence from proprietary control, it is often secondary to the urgent demands of speed. AIV ecosystems operate in safety-critical environments where vehicles must exchange and validate data in milliseconds to coordinate maneuvers and avoid collisions. Consequently, the AIV trilemma becomes a trade-off among Security, Speed, and Scalability [20]. Security ensures cryptographic integrity of vehicular data, resilience to replay and Sybil attacks, and trust in V2V and V2I communications. Speed guarantees real-time or near-real-time consensus and rapid data propagation to support cooperative driving. Scalability supports thousands of AIVs operating simultaneously in dense traffic environments with high-frequency event logging [21]. Numerous innovations have attempted to reconcile this trilemma. Sharding partitions the blockchain to enhance scalability but can weaken security or decentralization. Off-chain transactions, such as those supported by the Lightning Network, increase throughput but introduce new trust assumptions that may compromise security. Novel consensus algorithms, such as Proof-of-Stake, promise improvements in scalability and energy efficiency but risk altering decentralization dynamics [3]. Despite these advances, the trilemma persists as a pivotal barrier in deploying blockchain for AIVs, where safety-critical responsiveness and large-scale coordination cannot tolerate prolonged trade-offs. Thus, in AIV contexts, the blockchain trilemma is more than a theoretical limitation. It is a practical engineering obstacle. Unlike in finance or enterprise systems, where reduced throughput or partial centralization may be acceptable, in vehicular systems these trade-offs translate directly into risks for safety, reliability, and trust. Table 1 compares the classical blockchain trilemma with its AIV-adapted version, highlighting how the priorities shift in autonomous vehicular networks. While the classical model emphasizes decentralization, security, and scalability in generic terms, the AIV-adapted trilemma focuses on cryptographic data integrity, real-time consensus for safety-critical maneuvers, and scalability to handle dense, high-frequency vehicular communication.

2.1 Security Requirements in AIV

Autonomous Intelligent Vehicles rely on a sophisticated array of sensors, including LiDAR, radar, cameras, and ultrasonic devices, to perceive and interpret their surrounding environment. These sensors provide the critical inputs required for navigation, obstacle detection, and decision-making processes. The process of sensor fusion integrates these heterogeneous data streams into a coherent situational model, which is then used to guide actions such as lane changing, adaptive cruise control, and collision avoidance. The accuracy and trustworthiness of this fused data are therefore central to the safe operation of AIVs. However, the reliance on multi-sensor systems also introduces significant vulnerabilities. Autonomous vehicles are susceptible to sensor manipulation attacks, including spoofing and jamming, which can mislead perception modules by presenting false or distorted information. Persistent manipulation or corruption of sensor data not only endangers immediate safety but also undermines public trust in the reliability of autonomous vehicles, thereby impeding their widespread adoption. Breaches of data integrity in AIVs can occur through various vectors. Unauthorized access to onboard data storage, interception of communication between vehicles and infrastructure, or exploitation of vulnerabilities in supporting cloud services all present opportunities for attackers. Techniques such as direct hacking into vehicle control units, intercepting V2V or V2I transmissions, or exploiting weakly protected roadside units can corrupt or delay critical data exchanges. The consequences of such breaches are multifaceted. Exposure of sensitive data, including passenger identity or location, poses serious privacy concerns and risks such as real-time tracking or identity theft. More critically, compromised sensor and navigation data can directly affect vehicle behavior, leading to erroneous maneuvers, accidents, or systemic operational failures. High-profile incidents of this nature could also trigger regulatory backlash and slow down industry-wide deployment. Safeguarding the integrity of AIV sensor data requires a multilayered cybersecurity strategy. This includes employing strong encryption to secure data at rest and in transit, robust authentication and access control mechanisms, intrusion detection systems to monitor for anomalous activity, and regular security audits to identify latent vulnerabilities. Data anonymization further reduces risks by minimizing the impact of potential breaches on individual privacy [22].

Blockchain technology enhances these protections by offering a distributed and immutable ledger that secures the provenance of sensor data. By recording transactions in an unalterable manner, blockchain ensures that all vehicle logs, sensor records, and incident data are permanent and resistant to tampering. This capability is especially valuable for verifying the authenticity of messages in V2V and V2I communication, where rapid authentication and message integrity are critical for safe navigation [23]. Blockchain's decentralized data storage also mitigates single points of failure, ensuring that vehicular information remains available even if specific nodes are compromised [24]. In addition, blockchain enables the integration of smart contracts to automate agreements relevant to AIV operations, such as insurance claims, liability assignment, or maintenance scheduling. By removing intermediaries, smart contracts reduce the risk of fraud and improve transparency in transactions [25]. Similarly, blockchain provides a robust foundation for identity management systems, ensuring that only authenticated vehicles, manufacturers, or infrastructure components participate in the network, thereby reducing the likelihood of impersonation attacks [26]. Beyond data exchange, blockchain also strengthens other aspects of trust in AIV ecosystems. Secure and verifiable dissemination of software updates ensures that vehicles only receive genuine patches, preventing the installation of malicious software. Furthermore, blockchain enhances the transparency of the AIV supply chain by certifying the authenticity of hardware and software components before integration. These features collectively reinforce both the safety of navigation and the long-term trust required for the adoption of autonomous transportation systems. In summary, ensuring data integrity for sensor fusion and navigation is both a technical and societal necessity. The consequences of compromised data extend from immediate safety failures to long-term erosion of public trust in AIVs. Blockchain provides a powerful set of tools to address these challenges, offering immutable data records, decentralized security, authenticated communications, and automated trust enforcement. While its integration must still balance the trilemma of security, speed, and scalability, blockchain remains a compelling foundation for protecting sensor integrity and advancing the safe deployment of autonomous vehicles.

In the AIV ecosystems, node authentication is one of the most critical requirements for securing vehicular communications. Vehicles and infrastructure units such as roadside units, traffic lights, and edge servers continuously exchange information on speed, position, route intentions, and traffic conditions to support cooperative driving and enhance traffic efficiency [27]. Ensuring that each participating node is authentic, trusted, and authorized is therefore essential to prevent malicious actors from infiltrating the network and injecting false or misleading information. Without strong authentication mechanisms, AIV networks are exposed to a range of attacks. Sybil attacks, where an adversary forges multiple fake identities to overwhelm the system, can distort traffic coordination and resource allocation. Replay attacks [28], in which outdated or previously valid messages are resent, can mislead vehicles into making unsafe decisions. Similarly, impersonation attacks allow a malicious entity to pose as a legitimate vehicle or infrastructure component, sending false safety messages such as fabricated accident warnings or traffic light signals [29].

Blockchain technology provides a robust foundation for decentralized node authentication. Each vehicle or infrastructure node can be assigned a unique cryptographic identity anchored in the blockchain. Transactions broadcast within the vehicular network are signed using these identities and validated through consensus, ensuring that only legitimate participants are able to contribute data. This prevents unauthorized nodes from entering the system while maintaining a tamper-resistant log of all authentication events. By eliminating reliance on centralized authorities, blockchain-based identity management also avoids single points of failure, which are common vulnerabilities in traditional certificate-based approaches [30]. However, integrating blockchain-based authentication into AIV networks must account for the trilemma trade-offs. High levels of security achieved through strong cryptographic validation and consensus protocols can introduce latency, which is problematic for real-time vehicular communication. Furthermore, as the number of vehicles and infrastructure units grows, scalability becomes a pressing issue. Validating each authentication transaction across thousands of nodes may overwhelm network resources if not carefully optimized.

Emerging strategies attempt to balance these demands. Lightweight cryptographic protocols [31], hierarchical blockchain architectures, and edge-assisted authentication models have been proposed to reduce overhead while retaining strong security guarantees [32]. Smart contracts can also automate access control, ensuring that only authorized vehicles or infrastructure nodes can exchange data, thereby reducing the risk of impersonation. While these innovations demonstrate promise, the tension between security, speed, and scalability remains unresolved, making node authentication one of the central challenges in blockchain-enabled AIV systems.

AIV systems must guarantee data integrity in ensuring accuracy and consistency in sensor fusion and navigation to prevent catastrophic misinterpretation. Node authentication is also required in both V2V and V2I exchanges to prevent impersonation attacks [33]. Protection against threats in mitigating replay attacks, Sybil attacks, and data tampering is essential for maintaining network robustness. Blockchain immutable, cryptographically enforced ledger meets the demands for data integrity and authentication by making tampering computationally infeasible, however, scalability challenges and energy consumption remain significant barriers in real-world deployments [34]. In distributed vehicular ecosystems, particularly those supporting V2V and V2I communications, malicious actors may exploit the open and dynamic nature of the network to launch replay attacks, Sybil attacks, and data tampering attempts. These threats represent some of the most critical challenges in securing Autonomous Intelligent Vehicle (AIV) operations because they directly target the authenticity and integrity of the data upon which safety-critical decisions are based [35].

Replay Attacks occur when an adversary captures valid communication packets (such as vehicle position updates, hazard warnings, or traffic light signals) and retransmits them at a later time. While the original message may have been legitimate, replaying it outside of its valid time frame can cause vehicles to take inappropriate actions. For example, a replayed braking signal from a nearby vehicle may trigger unnecessary evasive maneuvers, potentially causing congestion or accidents. Mitigation requires strict time stamping, nonce-based protocols, and cryptographic freshness checks to ensure that only current, valid messages are processed. Blockchain immutable ledger strengthens this by recording event timelines that are verifiable, preventing attackers from reintroducing outdated data without detection. Sybil Attacks involve an adversary creating multiple forged identities within the vehicular network, enabling them to flood consensus protocols or distort cooperative driving processes. In an AIV scenario, Sybil nodes could, for instance, generate fake congestion alerts to manipulate traffic flow or force vehicles into less efficient routes. Traditional certificate-based Public Key Infrastructures (PKIs) can limit Sybil attacks, but they require centralized authorities, which present scalability and single-point-of-failure issues. Blockchain mitigates Sybil risks through decentralized identity management, where each vehicle is tied to a unique, verifiable cryptographic identity. Furthermore, consensus protocols can be designed to limit voting power or transaction influence based on stake or reputation, thereby reducing the effectiveness of identity forgery [36].

Data Tampering refers to unauthorized modification of transmitted or stored data within AIV systems. Attackers may attempt to alter sensor readings, navigation logs, or inter-vehicular messages to mislead decision-making processes. Even minor manipulations, such as falsifying GPS coordinates, can have catastrophic consequences in high-speed environments. Blockchain addresses tampering risks by using cryptographic hashing and immutable records, ensuring that once data is committed to the ledger, it cannot be altered without consensus from the network. This guarantees data provenance and makes it computationally infeasible for attackers to retroactively modify logs of sensor readings or vehicular interactions [37]. These protections, however, reintroduce the tensions of the blockchain trilemma. High-security measures such as frequent hashing, consensus validation, and decentralized authentication add computational and communication overhead. This increases latency, potentially slowing down real-time responsiveness that is essential for collision avoidance or coordinated maneuvers. Similarly, as vehicular networks scale to thousands of participants, maintaining resistance against Sybil nodes and validating the integrity of large volumes of data stresses blockchain's scalability. Solutions such as lightweight consensus protocols, off-chain verification, and hierarchical blockchain structures are therefore being explored to maintain robust protection without sacrificing operational performance.

2.2 Speed (Latency) Requirements

The requirement for real-time or near-real-time consensus is fundamental to safe operations. Vehicles interact continuously in dynamic environments where decisions must be validated and shared within milliseconds, this is particularly crucial in scenarios such as collision avoidance and cooperative maneuvers. In these cases, even minor delays in communication or agreement can lead to unsafe actions, chain reactions of errors, and potentially catastrophic accidents. If consensus delays accumulate, coordination breaks down, leading to safety hazards. To address this gap, researchers and industry practitioners are exploring latency optimized consensus models [38]. One approach is edge-assisted consensus, where local roadside units (RSUs) validate transactions within a geographically bounded cluster before relaying results to the global blockchain [39]. This localization reduces communication overhead and accelerates decision-making while preserving the immutability and auditability of blockchain records. Another solution is the use of off-chain mechanisms such as state channels and sidechains, which enable vehicles to exchange rapid, temporary agreements that can be settled onto the blockchain later. Hybrid consensus models also show promise, for instance by combining PBFT for localized, high-speed decisions with PoS for long-term secure record keeping. However, these innovations do not fully eliminate the inherent tensions of the blockchain trilemma. Real-time consensus is indispensable for the safe operation of AIVs, but it starkly reveals the conflict between security, speed, and scalability. Blockchain provides mechanisms to enhance trust and integrity in vehicular communication, yet without significant innovation in consensus protocols and system architecture, achieving latency-compatible blockchain for AIVs remains one of the greatest barriers to widespread deployment. Block propagation delay refers to the time it takes for a newly proposed block to be transmitted, verified, and distributed across all participating nodes. Consensus time refers to the additional period required for network participants to agree on the validity of the block and commit it to the ledger. Both metrics are crucial because they determine how quickly a piece of information becomes globally visible and trusted within the vehicular network. In traditional blockchains, block propagation may take several seconds, and final consensus may require multiple confirmations over minutes. These delays are tolerable in financial applications, where transaction settlement can wait, but they are unacceptable in AIV environments, where vehicles must make decisions within tens of milliseconds. For example, if a vehicle suddenly brakes to avoid a pedestrian, other vehicles in proximity must be notified and authenticated almost instantly to coordinate safe responses. A propagation delay of even one or two seconds undermines the entire premise of real-time coordination and exposes vehicles to collisions. The physical dynamics of vehicular communication networks further complicate propagation. High mobility, frequent topology changes, and variable signal quality in vehicular ad hoc networks (VANETs) can increase latency and reduce reliability. A block may not propagate evenly across the network, creating inconsistent knowledge states where some vehicles act on outdated information. Such asynchrony poses a critical risk in scenarios like intersections, where collective coordination is required. Blockchain-based solutions to propagation and consensus limits focus on reducing latency while preserving trust guarantees. Techniques such as smaller block sizes and faster block intervals reduce transmission overhead, although they also raise challenges related to orphaned blocks and security vulnerabilities. Hierarchical blockchain structures, where local clusters achieve rapid local consensus before syncing with the global chain, can also reduce propagation bottlenecks. The off-chain and layer-2 mechanisms allow time-critical data to bypass the slower global chain, with settlement occurring later for auditability. Additionally, edge computing integration enables geographically localized consensus, cutting propagation distances and minimizing latency. Yet, these adaptations highlight the blockchain trilemma in practice, an attempt to scale by aggregating more transactions per block increases block size, which in turn raises propagation delay.

2.3 Scalability Requirements

Scalability is a fundamental requirement in AIV ecosystems because vehicular networks must accommodate large numbers of simultaneously connected vehicles, especially in dense urban environments. A single metropolitan area may contain tens of thousands of AIVs, each generating a continuous stream of data that must be authenticated, shared, and validated in real-time. These data streams include positional updates, velocity, braking events, sensor readings, and hazard alerts, all of which are critical for cooperative driving and traffic safety [40]. In traditional blockchain platforms, scalability has long been a bottleneck. For instance, legacy networks such as Bitcoin and Ethereum support only a handful to a few dozen transactions per second, whereas vehicular environments may demand thousands of transactions per second per square kilometer. This discrepancy highlights the gap between the native capacity of blockchain and the operational requirements of intelligent transportation. If a blockchain-based vehicular system cannot support the required scale, it risks congestion, delayed validations, and network partitioning. Any such performance degradation would compromise safety, since delays in propagating verified information can cause vehicles to make decisions based on outdated or incomplete data [41]. Proposed solutions include sharding, where the network is divided into multiple smaller groups that process transactions in parallel, and clustered consensus architectures, where groups of vehicles or roadside units form localized sub-blockchains. These strategies reduce the burden on the global network by confining consensus to smaller domains while still preserving integrity through periodic synchronization. Edge computing integration further enhances scalability by offloading validation tasks to nearby roadside infrastructure, reducing latency and minimizing the amount of information that must traverse the entire blockchain network [42].

In addition to supporting large numbers of vehicles, blockchain-based vehicular systems must also manage the intensity of high-frequency event logging. AIVs generate enormous amounts of data every second, including sensor fusion outputs, navigation decisions, collision warnings, and cooperative intent signals. Recording all of these events on-chain would place unsustainable stress on blockchain throughput, storage, and communication bandwidth [43]. High-frequency event logging creates two major challenges. It accelerates ledger growth, since every transaction must be stored and verified by all nodes. It also increases network propagation load, as each event must be broadcast to maintain consistency [44]. To mitigate these issues, researchers propose hierarchical logging models and off-chain storage mechanisms. For example, only critical safety-related events may be recorded directly on-chain, while less critical, high volume data such as continuous sensor readings can be logged off-chain in distributed storage systems and anchored periodically through hash commitments.

This approach reduces blockchain congestion while ensuring that safety-critical information remains immutable and auditable. Layer-2 solutions, such as state channels, also provide a way to handle high frequency interactions locally between vehicles, with only final states committed to the blockchain [45]. Despite these strategies, scalability remains constrained by the blockchain trilemma. Optimizing for throughput by reducing on-chain data increases reliance on secondary storage and off-chain trust mechanisms, potentially weakening security. Conversely, prioritizing immutability and full on-chain validation slows down transaction rates, limiting the ability to keep pace with real-time vehicular data flows. Achieving scalability in the face of high-frequency logging without undermining speed or security continues to be one of the most pressing challenges for blockchain adoption in AIVs.

3 Innovation Design Intelligence Techniques for Secured AIV

The discussion of the blockchain trilemma in AIV highlighted the fundamental tensions between security, speed, and scalability [46]. While these trade-offs present significant barriers, research and industry practice have explored innovative techniques that combine blockchain with complementary technologies such as artificial intelligence (AI), edge computing, and smart contracts to design security frameworks tailored for autonomous transportation [47]. This section examines these design intelligence techniques, focusing on how they mitigate trilemma constraints while enabling practical, real-world deployment of blockchain in AIV ecosystems.

3.1 Hybrid Consensus Mechanisms

Hybrid consensus mechanisms have emerged as a promising solution to reconcile the blockchain trilemma in AIV systems. A particularly effective model combines Practical Byzantine Fault Tolerance (PBFT) for low latency operations within local clusters, with Proof-of-Stake (PoS) or Proof-of-Authority (PoA) algorithms for global state synchronization [48]. This hybrid enables real-time responsiveness for critical safety functions, while ensuring secure and scalable network-wide consistency. PBFT is suitable for permissioned environments, such as a localized team of nearby vehicles or edge nodes, where fast and deterministic consensus is crucial. Its design ensures safety as long as no more than one-third of nodes are malicious, and it achieves consensus in a fixed number of communication rounds. This enables millisecond-level agreement, which is ideal for latency-critical tasks such as collision avoidance or platooning [49]. However, PBFT's communication complexity ($O(n^2)$) limits its scalability when deployed across large networks. Its performance typically degrades when the number of nodes exceeds a few hundred [50].

Table 2 Comparison of consensus mechanisms in hybrid blockchain for AIVs

Layer	Consensus mechanism	Strengths	Trade-offs managed
Local	PBFT	Ultra-low latency, deterministic finality, suitable for small clusters	Limited scalability due to $O(n^2)$ communication complexity
Global	PoS/PoA	Scalable, efficient, energy-saving, suitable for high-volume environments	Higher latency than PBFT, reduced decentralization (PoA)
Hybrid	PBFT + PoS/PoA	Balances security, speed, and scalability by combining local responsiveness with global scalability	Requires careful synchronization design, added system complexity

At a larger scale, across entire urban networks or multi-city deployments, blockchains must process high volumes of data securely and efficiently. Proof-of-Stake (PoS) or Proof-of-Authority (PoA) consensus mechanisms provide effective solutions for node trustworthiness through staking rather than identity. PoA further improves performance by limiting validation rights to authorized entities, reducing overhead at the cost of some decentralization. PoS and PoA are better suited to ensure scalability and energy efficiency while still maintaining acceptable security standards [51]. In an AIV context, the hybrid strategy may function as a local cluster consensus and global chain synchronization. In local cluster consensus, vehicles in proximity validate critical events using PBFT among a small cluster, ensuring rapid agreement within local latency requirements [52]. In the global chain synchronization, cluster summaries or state snapshots are committed to the global blockchain via PoS or PoA consensus periodically or as needed, allowing for network wide consistency handling large scale data aggregation effectively. This hybrid consensus strategy mitigates the trade-off as PBFT supplies speed and security locally, PoS/PoA provides scalability and continued trust at the global level [53, 54]. Table 2 presents a comparison of consensus mechanisms in hybrid blockchain for AIVs, highlighting the trade-offs between local, global, and hybrid approaches. It shows how PBFT supports ultra-low latency at the local level, PoS/PoA ensures scalability at the global level, and a hybrid model balances security, speed, and scalability, albeit with added system complexity.

3.2 Edge-Blockchain Architecture

One of the most significant innovations in addressing the blockchain trilemma for AIVs is the integration of edge computing with blockchain, commonly referred to as an edge–blockchain architecture. This design distributes computation and consensus closer to the data source, such as vehicles and roadside units (RSUs), thereby reducing latency, improving scalability, and preserving data security. In traditional blockchain networks, all transactions must be validated and propagated across the entire network, introducing propagation delays that are incompatible with the real-time requirements of AIVs. By integrating edge nodes, computational and consensus tasks are executed at geographically closer locations, enabling localized consensus [55]. For instance, RSUs positioned along highways or in intersections can serve as validators, ensuring that safety-critical events such as sudden braking, collision alerts, or lane-change signals are authenticated within milliseconds. Studies demonstrate that edge-assisted blockchain systems in vehicular environments can reduce consensus latency by up to 40% compared to centralized cloud-based blockchains [56, 57].

Beyond latency improvements, edge blockchain architectures address scalability challenges by confining high frequency transactions to localized sub-chains, which periodically synchronize with the global blockchain. This prevents network congestion caused by continuous logging of vehicular events on a single global ledger. In dense urban environments, where thousands of AIVs may operate simultaneously, edge blockchain systems enable parallel processing and hierarchical validation, significantly improving throughput. Research on edge-enabled blockchain for intelligent transportation has shown that hierarchical frameworks can support thousands of concurrent nodes with reduced communication overhead [58]. Integrating blockchain with edge computing also strengthens trust and security. Blockchain ensures data integrity and immutability, while edge servers provide localized intrusion detection and anomaly monitoring. This dual mechanism allows the system to filter malicious data before it spreads through the global network. Furthermore, decentralized data storage at the edge prevents single points of failure, ensuring network resilience even if some nodes are compromised [59]. In summary, practical applications of edge blockchain architectures in AIVs include the following;

- V2V and V2I authentication using local edge validators to ensure secure data exchange.
- Real-time traffic management, where RSUs aggregate vehicular data, validate transactions, and forward summaries to global chains.
- Incident reporting and liability management, in which local sub-chains instantly record collision data and synchronize it with city-level blockchains for insurance or regulatory purposes.
- Efficient software update distribution, where updates are validated and disseminated via edge nodes to reduce load on the central infrastructure.

Using federated blockchain tiers which distribute blockchain functionality across edge and cloud layers has been a promising architecture that leverages localized responsiveness at the edge while using cloud infrastructure for scalability and historical record management [60]. By structuring blockchain operations into distinct tiers, the system addresses both the speed and scalability challenges of AIV networks without sacrificing security. While the edge layer ensures immediacy, the cloud layer provides the scalability and storage needed for long-term record keeping. Data that are less time-sensitive, such as driving history, traffic patterns, maintenance logs, and liability records, are aggregated and periodically synchronized with a cloud-hosted blockchain. This layer can employ scalable consensus protocols such as PoS or PoA, which handle large-scale data aggregation efficiently.

The cloud layer acts as the system of record, preserving the immutability and auditability of historical data while offloading the edge layer from the burden of long-term storage. In practice, this duality allows the blockchain to serve both real-time operational needs and regulatory or forensic requirements. For example, collision data validated at the edge can later be anchored in the cloud blockchain to support insurance claims, legal accountability, or fleet management analysis. This federated design also enhances security. The edge layer mitigates risks of real-time tampering by immediately validating data at the source, while the cloud layer guarantees that all records are immutable and auditable over the long term [61]. Decentralized storage across both tiers eliminates single points of failure and provides redundancy in case of localized disruptions. Furthermore, smart contracts deployed across tiers can automate cross-layer interactions, such as syncing edge summaries with cloud ledgers or initiating automatic reporting after accidents.

Figure 1 is the flow diagram of the edge blockchain architecture, where the vehicles at the bottom layer continuously generate sensor data and broadcast critical safety events. Then at the edge layer, where PBFT clusters operate, nearby vehicles and RSUs form small-scale blockchain clusters for real-time consensus, ensuring low latency in decision-making. These aggregated data from edge clusters is periodically synced to the cloud for long-term storage, analytics, and regulatory compliance. This federated design balances the blockchain trilemma with respect to speed, scalability, and security.

3.3 AI-Augmented Security

Using machine learning (ML) and deep learning (DL) algorithms, AIV ecosystems can strengthen the blockchain security layer in two main ways: anomaly detection in transaction patterns and predictive optimization of validation selection. One of the key threats in blockchain-enabled vehicular networks is the presence of malicious nodes that attempt to launch attacks such as Sybil, replay, or false data injection. Blockchain alone provides immutability and transparency but lacks the ability to proactively detect such threats in real-time. AI-based anomaly detection addresses this gap by analyzing transaction patterns, network behavior, and communication flows to identify abnormal activities before they propagate through the

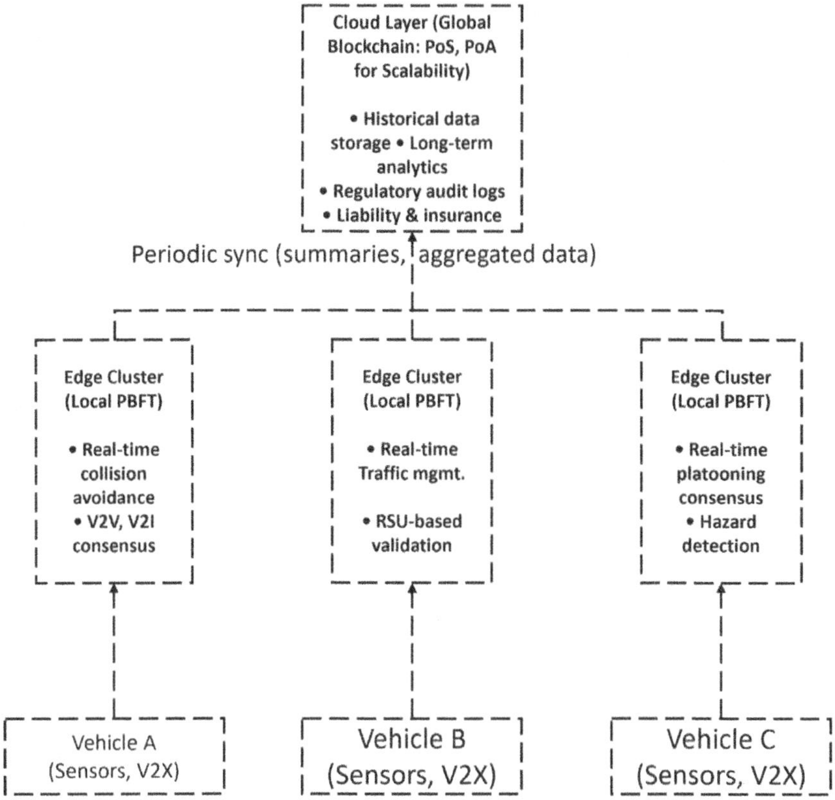

Fig. 1 Federated edge cloud blockchain architecture showing how the edge layer and cloud layer interact for secure AIV operations

ledger. Machine learning models, including Support Vector Machines (SVMs), Random Forests, and Deep Neural Networks (DNNs), have been applied in blockchain-driven VANETs to identify suspicious transactions with high accuracy [62]. For example, anomaly detection frameworks that combine blockchain with AI demonstrated improved detection rates of Sybil and DoS attacks in vehicular networks, reducing false positives compared to rule-based systems [63, 64]. Furthermore, deep learning-based intrusion detection integrated with blockchain log analysis has been shown to achieve near real-time detection of anomalies, enabling proactive defense in mobility networks [65]. This approach ensures that AIVs are not only operating in an immutable ledger environment but also protected by an intelligent detection layer capable of learning from evolving threat behaviors. Another challenge in blockchain-enabled AIV systems is the selection of validators, which determines which nodes participate in consensus. Traditional consensus algorithms, such as PoS or PBFT, may assign validators based on stake or pre-assigned authority, but they do not dynamically account for network state, trust level, or node reliability. AI introduces

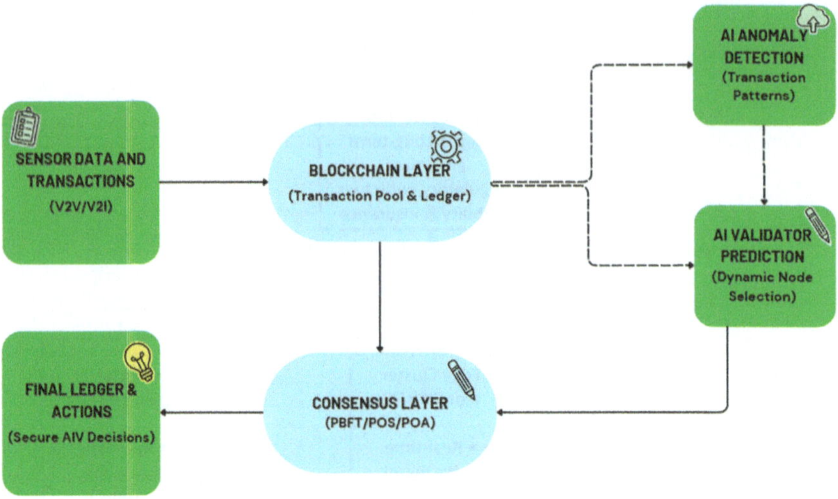

Fig. 2 AI-based modules integrating into blockchain pipelines for AIV security

predictive models that evaluate historical trust scores, network latency, and resource availability to optimize the selection of validators [66]. Using reinforcement learning and predictive analytics, validators can be dynamically chosen to balance security and performance [67]. For example, AI-based trust management systems have been integrated with vehicular blockchain frameworks to select the most reliable nodes for consensus, improving both throughput and resistance to malicious validators [68]. In AIV contexts, AI-driven anomaly detection strengthens resilience against malicious activities beyond cryptographic guarantees, predictive validator selection reduces delays by assigning consensus tasks to the most optimal nodes, and AI models handle large volumes of blockchain data and adapt consensus decisions dynamically, enabling the system to grow without exponential overhead. This means that low-latency edge nodes can be prioritized during high speed collision avoidance scenarios, while more resource-rich nodes are selected for global consensus, ensuring that consensus remains adaptive and resilient, directly mitigating the blockchain trilemma trade-offs.

Figure 2 shows the AI augmented blockchain system showing how sensor data and transactions from AIVs enter the blockchain layer while AI modules enhance security. Anomaly detection scans for malicious patterns, while validation model prediction selects reliable nodes for consensus. These inputs are fed into the consensus layer PBFT/PoS/PoA, and the validated results are stored in the final ledger, enabling secure and trustworthy AIV decision-making. The output is committed to the Final Ledger and Actions.

3.4 Lightweight Cryptography

A critical requirement for blockchain-enabled AIV systems is the use of lightweight cryptographic algorithms that balance strong security with the limited computational and energy resources available in embedded vehicular hardware [69]. Unlike traditional IT systems, AIVs rely on electronic control units (ECUs), sensors, and communication modules with limited processing power and memory. Elliptic-Curve Cryptography (ECC) has become a cornerstone of lightweight cryptography in vehicular networks. Compared to RSA, ECC provides equivalent security with much smaller key sizes, for example, a 256-bit ECC key offers similar security to a 3072-bit RSA key [70]. This reduction directly results in faster computations, reduced bandwidth usage, and lower storage overhead, which are vital for real-time V2V and V2I communications. For blockchain-enabled AIVs, ECC can be integrated into transaction signing, certificate management, and consensus validation without imposing prohibitive computational delays. Although ECC is efficient for current vehicular systems, the advent of quantum computing poses a long-term risk, as algorithms such as Shor's could break ECC and RSA. Consequently, Post-Quantum Cryptography (PQC) is being actively studied for vehicular blockchain applications [71]. Lightweight PQC algorithms such as CRYSTALS-Kyber and CRYSTALS-Dilithium, both standardized by NIST in 2023, offer quantum-resistant security while being optimized for embedded devices [72]. Recent studies show that lattice-based PQC algorithms can be adapted to run on automotive grade micro controllers with feasible computational and memory footprints, enabling them to protect AIV communications against quantum-capable adversaries [73]. Implementing cryptographic solutions that are both secure and efficient is therefore essential to ensure real-time responsiveness, low energy consumption, and resilience against emerging threats, including quantum attacks [74]. Lightweight cryptography directly contributes to resolving trilemma constraints in AIVs in that ECC and PQC provide robust protection against classical and quantum threats, reduced key sizes, and optimized algorithms enable low-latency authentication and consensus and efficient use of computational and communication resources ensures that cryptographic overhead does not bottleneck large-scale vehicular deployments.

Figure 3 shows a visual representation of the blockchain trilemma in AIVs, highlighting the trade-offs among security, speed, and scalability. The triangular framework illustrates how improving one dimension often impacts the others, security ensures protection against cyberattacks and data tampering, speed reflects the low-latency requirements of safety-critical maneuvers, and scalability captures the ability to support thousands of vehicles in dense traffic. The central balance icon emphasizes the challenge of simultaneously optimizing all three dimensions, while the surrounding icons illustrate technologies such as AI-driven validation, lightweight cryptography, and edge-based blockchain architectures that can help manage these trade-offs. In summary, these innovation driven strategies demonstrate how the blockchain trilemma in AIVs can be managed through complementary approaches rather than a single solution. Hybrid consensus mechanisms enhance speed by enabling sub-200

Fig. 3 Blockchain trilemma in AIVs, highlighting the trade-offs among security, speed, and scalability

Table 3 Innovation Strategies and Trilemma Dimension Focus

Strategy	Trilemma dimension	Contribution
PBFT (local clusters)	Speed	Sub-200 ms latency for safety-critical consensus
PoS/PoA (global sync)	Scalability	Efficient validation at scale with reduced overhead
Edge–blockchain tiers	Speed + Scalability	Local consensus at RSUs, cloud sync for large-scale record keeping
AI-based anomaly detection	Security	Identifies Sybil, replay, and false data injection attacks
AI validator prediction	Speed + Security	Selects reliable nodes, reduces consensus delay
Lightweight ECC/PQC	Security + Speed	Strong cryptography without overwhelming ECUs

ms decisions at the local level while maintaining security and consistency globally. Edge-blockchain architectures reduce propagation delays, supporting real-time responsiveness while scaling across city-wide deployments, thus directly addressing scalability. AI-augmented security modules add adaptability by strengthening security against evolving threats while simultaneously improving speed through predictive validator selection. Lightweight cryptography ensures that strong security can be achieved without overwhelming embedded vehicular hardware, thereby sustaining both speed and scalability in dense traffic conditions. Table 3 summarizes how each proposed innovation directly mitigates one or more dimensions of the blockchain trilemma in intelligent vehicular networks. Together, these innovations highlight that the trilemma can be balanced through layered architectures and adaptive designs, paving the way for practical blockchain applications in safety-critical vehicular ecosystems.

4 Practical Blockchain Applications for Security and Connectivity in AIV

To optimize blockchain security and connectivity in AIV ecosystems, several strategies have been proposed that directly address the challenges of the blockchain trilemma. One approach is the adoption of permissioned decentralized architectures with tiered access, where authenticated entities such as vehicles, roadside units, and manufacturers participate with defined roles, thereby reducing attack surfaces and latency while ensuring trustworthy communication [75]. A second strategy focuses on federated and hybrid connectivity models, where blockchain is integrated with federated learning and cluster-based consensus to support both local responsiveness and global synchronization [76, 77]. Reliable connectivity also depends on robust V2V and V2I standards and blockchain-based identity management to ensure interoperability and privacy in cross-manufacturer networks. At the consensus layer, adaptive and lightweight consensus mechanisms can dynamically adjust block sizes, validator sets, and intervals to maintain performance under varying network conditions [78]. Another important strategy is data minimization and event prioritization, whereby only safety-critical events (e.g., braking, hazard alerts) are recorded on-chain while less urgent or voluminous data is logged off-chain with on-chain hashes for auditability [79]. Finally, resilience and redundancy mechanisms are crucial: redundant roadside units, fallback communication paths, tamper-evident forensics, and secure OTA updates provide continuity and recovery in the event of attacks or failures [80]. Collectively, these strategies improve security through cryptographic trust and resilient design, improve speed through localized and adaptive consensus, and support scalability through hierarchical architectures, event prioritization, and standardized communication protocols. In a broader sense, we discuss these strategies in the context of security, speed, scalability, and resilient fault recovery.

4.1 Security-First Strategies

In the context of blockchain-enabled Autonomous Intelligent Vehicle (AIV) systems, security must be prioritized as the foundation for reliable connectivity and decision-making. Since AIVs operate in safety-critical environments, even minor compromises in identity management or data integrity can have severe consequences. Thus, security first strategies form the baseline of any blockchain deployment for intelligent transportation. Two major approaches are central: multi-factor node authentication and end-to-end encrypted communication channels.

4.1.1 Multi-factor Node Authentication

Blockchain enables decentralized identity verification in vehicular communication, ensuring that cryptographic identities are tamper-proof and preventing Sybil or impersonation attacks by anchoring each node credentials in an immutable ledger. Conventional public key infrastructures (PKI) provide cryptographic keys for authentication, but in highly dynamic vehicular environments, this may not be sufficient. To address this, multi-factor node authentication combines cryptographic keys with hardware security modules (HSMs) or Trusted Platform Modules (TPMs) embedded within vehicles and roadside units. This ensures that even if a key is compromised, authentication cannot proceed without validation from the hardware component. Research shows that blockchain-enhanced authentication schemes leveraging HSMs reduce the risk of identity forgery and provide stronger guarantees of trust in V2V and V2I communications [81].

4.1.2 Encrypted V2V and V2I Channels with Blockchain Integration

Blockchain transaction signing integrates directly with encrypted channels, ensuring that all V2V and V2I exchanges are not only confidential but also immutable and verifiable within the ledger. AIVs exchange high frequency, time sensitive data, such as collision alerts, braking signals, or traffic updates that must remain confidential and tamper-resistant. End-to-end encryption ensures confidentiality, while integrating blockchain transaction signing into V2V and V2I exchanges guarantees immutability and accountability. Messages are both encrypted for privacy and signed as blockchain transactions for non-repudiation. This dual-layer approach allows real-time communications to remain secure while providing an immutable audit trail for forensic or regulatory purposes. Studies highlight that blockchain-enhanced encryption frameworks significantly improve resilience against man-in-the-middle and replay attacks in vehicular adhoc networks [82, 83]. These security-first strategies directly reinforce the security pillar of the blockchain trilemma while complementing speed and scalability goals. Multi-factor authentication ensures that only verified nodes participate, reducing malicious overhead in consensus and preserving transaction throughput. Encrypted channels integrated with blockchain signatures reduce the risk of data manipulation without compromising communication latency. While these measures introduce some computational overhead, lightweight cryptographic protocols and hardware-assisted modules mitigate performance trade-offs, enabling security to remain the primary enabler of trustworthy AIV connectivity.

Figure 4 shows how authentication and encryption work together with blockchain signing to ensure secure AIV connectivity. Nodes initiate communication and undergo multi-factor authentication as messages pass through encrypted V2V/V2I

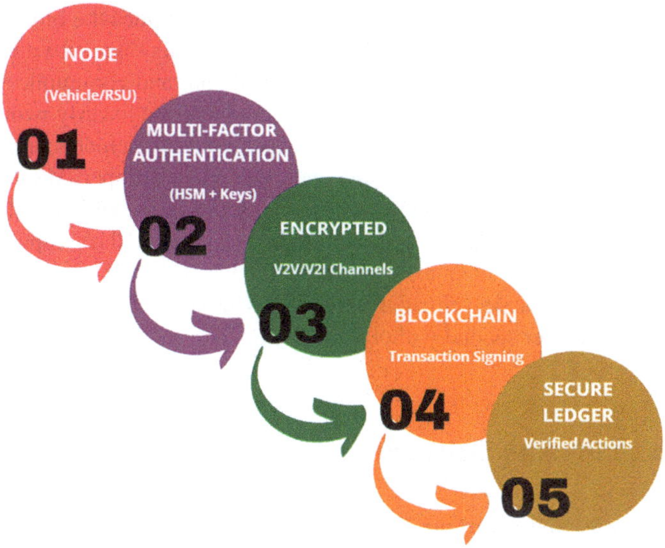

Fig. 4 Security-First Strategies in blockchain-enabled AIVs

channels. Transactions are then signed on the blockchain as data is committed to the secure ledger for verified actions.

4.2 *Speed Optimization Strategies*

After ensuring that only trusted participants communicate securely, the next challenge is to ensure that those secure transactions happen fast enough for AIV safety. Two prominent approaches are the adoption of Layer-2 solutions and sharding mechanisms, both of which aim to reduce latency and improve throughput without undermining security. Layer-2 technologies allow vehicles to conduct rapid, off-chain interactions while using the main blockchain only for settlement. State channels enable a set of participants to exchange transactions directly and instantaneously, with only the final state anchored to the blockchain. This drastically reduces on-chain congestion and supports micro-transactions, such as continuous updates of vehicle positions or cooperative maneuver agreements. Side chains, on the other hand, operate as auxiliary blockchains connected to the main chain, where specific vehicular tasks can be processed independently, providing the real-time responsiveness required for vehicular networks, while still preserving the immutability of the blockchain for final settlement and accountability [84, 85].

Sharding is another speed optimization strategy that divides the blockchain into multiple partitions known as "shards", each processing its own subset of transactions in parallel. This design reduces the computational and communication overhead on individual nodes while allowing throughput to scale linearly with the number of shards. In AIV contexts, sharding enables clusters of vehicles or geographic regions to process their local transactions independently, with periodic cross-shard synchronization to maintain global consistency. Research on sharded blockchains like Ethereum 2.0 and PolyShard demonstrates significant throughput improvements while maintaining adequate security guarantees [86, 87]. For AIV systems, this means that large-scale vehicular networks in urban environments can handle thousands of events per second without bottlenecks.

4.3 Scalability Enhancement

Although security and speed are essential for safe operation of AIVs, scalability determines whether blockchain solutions can realistically support dense urban environments with thousands of vehicles and high-frequency communication events. Without effective scaling strategies, blockchain networks are prone to congestion, poor performance, and excessive energy or bandwidth consumption. Two promising approaches to scalability in vehicular blockchain networks are cluster-based consensus and hierarchical blockchain architectures. In traditional blockchain systems, all nodes participate in validating every transaction, which can overwhelm networks with high transaction frequency, such as AIV ecosystems. Cluster-based consensus alleviates this issue by organizing vehicles and roadside units (RSUs) into local committees or clusters. Within each cluster, a lightweight consensus protocol, such as Practical Byzantine Fault Tolerance (PBFT), validates transactions locally. Only the aggregated or summarized results are then transmitted to higher levels of the blockchain network. This approach significantly reduces global communication overhead, while still ensuring data reliability at the local scale. For example, nearby vehicles coordinating a platoon or sharing lane change alerts can reach fast local consensus without burdening the entire city-wide ledger. Studies on committee based consensus models demonstrate improved throughput and reduced propagation delays compared to fully replicated global consensus mechanisms [50, 88]. To further enhance scalability, hierarchical blockchain architectures divide the vehicular network into multiple levels, for example, urban, regional, and global level blockchains. At the bottom tier, local ledgers capture high frequency safety-critical events within small geographic zones such as an intersection or highway segment. Periodically, these local chains are synchronized with regional blockchains, which aggregate information across larger areas. At the highest tier, global or national blockchains store summaries for long-term accountability, regulatory oversight, and inter-regional interoperability. This layered design ensures that local traffic does not overwhelm the global network while maintaining end-to-end transparency and traceability. In practice, hierarchical

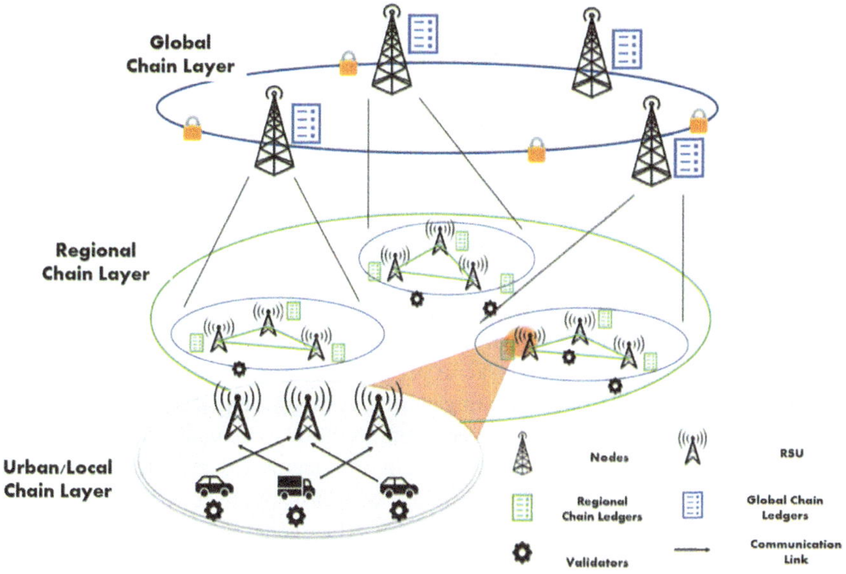

Fig. 5 Blockchain networks for scalability in AIVs showing urban/local blockchain, regional blockchain and global blockchain

models support both real-time vehicular decision-making and large-scale data management for transportation authorities. Research on multi-level blockchain frameworks for intelligent transportation confirms their potential to reduce bandwidth usage and increase transaction throughput in smart mobility systems [89].

Figure 5 shows how scalability is achieved. Local chains keep latency low and handle real-time safety-critical events at intersections or highways, regional chains manage city-wide traffic, aggregates, and synchronizes data across a city or metropolitan area, and the global chain ensures overall integrity, maintains inter-regional data consistency, and regulatory oversight.

4.4 Resilience and Fault Recovery

Blockchain redundancy and immutable records inherently support forensic fault recovery, even if some validators fail or data streams are interrupted, the distributed ledger preserves an untampered record for post-event validation and accountability. Unlike conventional distributed systems, AIVs function in dynamic and safety-critical environments where redundancy and fault recovery mechanisms must be tightly integrated to ensure continuity of service. One strategy for enhancing resilience is the deployment of redundant validators in zones characterized by high vehicular density, such as urban intersections or highways. In such contexts, blockchain-based consensus requires rapid validation of safety-critical messages. By having multiple validators operating in parallel within the same geographic cluster, the system ensures that consensus can still be achieved even if one or more validators fail or are compromised. Redundancy improves fault tolerance by distributing workload across multiple nodes and prevents single points of failure. Studies on vehicular blockchain frameworks highlight that validator redundancy reduces the probability of service disruptions and strengthens trust in dynamic environments [90, 91].

Another essential resilience mechanism is automatic failover in consensus nodes. In traditional blockchain networks, if validator nodes crash or lose connectivity, consensus processes stall until nodes recover or replacements are introduced manually. In AIV contexts, such delays are unacceptable. Automatic failover mechanisms dynamically reassign validation tasks to backup nodes or alternative clusters when primary validators fail. By incorporating predictive monitoring, the system can detect potential node failures early and trigger reconfiguration to maintain continuous consensus. Research in fault-tolerant blockchain architectures demonstrates that such mechanisms improve availability and reduce consensus recovery time to near real-time, making them suitable for safety-critical AIV scenarios [92]. Resilience and fault recovery strategies directly enhance the security dimension of the blockchain trilemma while indirectly supporting speed and scalability. Redundant validators ensure that transaction validation continues without latency spikes during node failures, while automatic failover mechanisms preserve throughput and prevent service degradation under adverse conditions. Together, these mechanisms establish the robustness required for blockchain adoption in AIV ecosystems, ensuring that vehicular networks can withstand both accidental failures and malicious disruptions.

5 Case Study

5.1 Case Study I: Urban Intersections

Consider a busy urban intersection during rush hour where more than 200 AIVs converge simultaneously. Each vehicle continuously exchanges critical safety data, such as braking events, hazard detection alerts, and cooperative lane change intentions with nearby vehicles (V2V) and road infrastructure (V2I). Managing this level of high frequency communication while preserving security, speed, and scalability poses a major challenge under the blockchain trilemma.

5.1.1 Tiered Blockchain with AI-Driven Consensus

In this scenario, a tiered blockchain architecture is deployed. At the lowest tier, a local blockchain cluster formed by nearby vehicles and roadside units validates transactions using a low-latency consensus protocol, such as PBFT [93]. This allows vehicles to agree on collision avoidance maneuvers and cooperative braking within 200 ms, ensuring that safety-critical decisions are executed in real-time. Periodically, aggregated transaction summaries are uploaded to a regional blockchain, which manages coordination throughout the city. Finally, a global blockchain records longer term summaries for regulatory oversight and liability resolution, maintaining immutability and transparency across jurisdictions, as demonstrated in hierarchical mobility frameworks proposed by [94].

5.2 AI-Augmented Decision-Making

The consensus process is further enhanced by AI-driven modules. Anomaly detection system inspects transaction patterns in real-time to identify potential malicious nodes, filtering out false braking signals or injected hazard alerts. At the same time, an AI-based validator prediction system dynamically selects the most reliable vehicles and RSUs to participate in consensus, balancing latency and trustworthiness [56]. Performance Outcomes Simulation and empirical studies demonstrate that blockchain-augmented vehicular systems, when combined with edge computing, can maintain sub-200 ms consensus latency, a threshold essential for collision avoidance and cooperative maneuvers [95]. In addition, frameworks integrating blockchain and edge layers have shown the ability to handle large scale vehicular transactions while ensuring immutability and trust [58]. In parallel, advanced sharding-based approaches such as Kronos [96] and GriDB [97] illustrate pathways for scaling blockchain throughput to thousands of transactions per second, which is critical for dense intersections with hundreds of AIVs. This case highlights how combining tiered blockchain architectures with AI-driven consensus optimization directly addresses the blockchain trilemma in AIV contexts. Security is reinforced through anomaly detection and immutable records; speed is achieved via localized, low-latency consensus; and scalability is ensured through hierarchical blockchain tiers and sharding mechanisms that prevent global network congestion. Together, these mechanisms create a trustworthy digital backbone for safe and efficient AIV operations in complex, high-density urban environments.

5.3 Case Study II: Highway Platooning

A different but equally demanding scenario for blockchain in AIVs occurs in highway platooning, where groups of autonomous vehicles travel in tightly coordinated formations at speeds exceeding 100 km/h. In such settings, maintaining low latency communication, ensuring data integrity, and enabling scalability across hundreds of vehicles are critical for avoiding accidents and optimizing traffic flow. Each vehicle must continuously exchange telemetry data including speed, acceleration, braking force, and lane position with other platoon members, as well as with road infrastructure for coordination.

5.3.1 Blockchain-Enabled Platoon Coordination

In this environment, the blockchain provides a trust framework for secure communication among platoon vehicles. Transactions such as braking commands or lane-change intentions are recorded on a local blockchain tier, where cluster-based consensus validates the data in near real-time. This ensures that no malicious or faulty nodes can inject false signals into the platoon, which could otherwise cause cascading collisions. As demonstrated in committee-based vehicular blockchain systems [98], cluster-level consensus reduces communication overhead and accelerates decision-making in localized vehicular groups.

5.3.2 AI-Augmented Consensus for Predictive Safety

Beyond validation, AI-driven consensus optimization strengthens platoon safety by predicting anomalous behaviors and pre-selecting trustworthy validators. An anomaly detection module monitors braking and acceleration patterns to identify potentially compromised vehicles, while a predictive validator selection module ensures that only reliable vehicles participate in consensus. These approaches, highlighted by [99, 100], reduce latency while maintaining resilience against cyber attacks.

5.3.3 Multi-tier Blockchain for Scalability

Given that highway platoons often integrate into larger regional traffic systems, the blockchain must scale to support thousands of vehicles. Here, a multi-tier blockchain architecture is applied. Platoon-level blockchains handle local real-time consensus, regional blockchains synchronize aggregated platoon data, and a global blockchain ensures transparency for regulators and insurers. Reference [101] demonstrate that hierarchical models of this kind significantly reduce bandwidth usage while supporting real-time vehicular transactions. Simulation-based studies confirm that blockchain frameworks augmented with edge computing can achieve consensus within 100–150 ms in vehicular environments [102]. This latency is well within the safety thresholds required for platooning maneuvers, such as coordinated braking at highway speeds. Additionally, sharding protocols provide scalable solutions for managing the large volumes of event data generated across multiple platoons simultaneously. By integrating lightweight local consensus, predictive validator selection, and hierarchical synchronization, AIV platoons can achieve the speed, security, and scalability balance required to safely operate in dynamic highway environments.

Table 4 Case studies: urban intersection verses highway platooning

Dimension	Case study 1: urban intersection	Case study 2: highway platooning
Context	200+ AIVs converging at a busy city intersection during rush hour. High-frequency transactions for braking, lane changes, and hazard alerts. Dense network with high communication overhead	Multiple AIVs traveling in tightly coordinated platoons at highway speeds (+=100 km/h). Continuous telemetry sharing: acceleration, braking, lane position. Latency-sensitive environment for safety
Blockchain Architecture	Tiered blockchain: Local intersection blockchain (fast PBFT consensus). Regional blockchain (aggregates urban data). Global blockchain (regulatory oversight)	Hierarchical blockchain: Platoon-level local chain (real-time safety validation). Regional chain (aggregates multiple platoons). Global chain (accountability and synchronization)
AI Integration	Anomaly detection filters malicious or false hazard/braking signals. Validator prediction ensures trustworthy nodes for local consensus	Anomaly detection monitors platoon telemetry for anomalies. Validator prediction selects reliable vehicles to reduce latency
Performance Target	Sub-200 ms latency for intersection-level consensus. Supports large volumes of event logs	100–150 ms latency for braking/lane-change coordination. Ensures safety at highway speeds
Scalability	Cluster-based consensus reduces global traffic. Hierarchical layers prevent city-wide congestion	Multi-tier scaling: platoon regional global. Sharding protocols support thousands of concurrent platoons
Key Benefits	Enhanced safety in dense traffic. Accountability for accident forensics. Efficient coordination at urban scale	Safer platoon maneuvers at high speed. Lower energy use via coordinated driving. Global accountability and trust

Security is reinforced by anomaly detection that prevents malicious braking or lane-change signals from destabilizing the platoon; Speed is safeguarded by consensus latencies of 100–150 ms, which fall within safe thresholds for high-speed vehicular coordination; and Scalability is achieved through hierarchical synchronization, enabling multiple platoons to operate concurrently without overwhelming regional or global networks. Table 4 compares these scenarios to illustrate how blockchain and AI can address distinct AIV challenges. It highlights differences in context, blockchain architecture, AI integration, performance targets, scalability approaches, and key benefits, showing how each scenario balances security, speed, and scalability under different traffic conditions.

6 Challenges and Future Research

Although blockchain offers significant potential to improve AIVs, the blockchain trilemma, balancing security, speed, and scalability, remains a defining obstacle. The challenges are not isolated technical issues, but interconnected constraints that ripple through system design, deployment, and long-term viability. The first critical limitation arises from the hardware and resource constraints of AIVs. Embedded vehicle systems such as ECUs and onboard sensors are optimized for perception and control rather than for executing resource intensive cryptographic operations. Maintaining even a partial ledger, performing frequent signature verifications, and participating in consensus protocols can overwhelm limited CPU, memory, and storage capacities. This resource tension is exacerbated when high throughput event logging is required in dense urban networks.

Beyond hardware, another pressing difficulty lies in the latency to security trade-off. Safety-critical maneuvers, such as brake or evasive lane changes, require decisions in 100–200 ms. Yet ensuring robust security requires multiple consensus rounds, cryptographic validation, and replay protection, all of which introduce propagation delays. As traffic density increases, so too does the volume of consensus traffic, compounding the risk that systems may fail to meet real-time deadlines. Thus, latency and security cannot be treated in isolation but must be carefully balanced within the broader design of vehicular blockchain protocols. This problem is further magnified by the scalability challenge. Dense traffic scenarios generate a torrent of V2V and V2I messages, creating throughput demands far exceeding the capacities of conventional blockchain networks, which typically manage only tens of transactions per second. The need to replicate or propagate transactions across thousands of vehicles quickly overwhelms both bandwidth and storage. Pruning or offloading data may ease resource pressure, but it risks undermining the immutability and forensic reliability that make blockchain attractive in the first place.

Energy efficiency introduces another layer of complexity. Consensus algorithms, even those less intensive than Proof-of-Work, place a significant burden on both computation and communication resources. For electric vehicles in particular, such overhead translates directly into reduced driving range and higher operating costs. Striking a balance between robust security and sustainable energy consumption remains a formidable challenge, as excessive optimization in either direction undermines overall system reliability. Finally, the issue of interoperability underscores the practical difficulty of integrating blockchain into real-world mobility systems. Current transport infrastructures, traffic lights, GPS services, highway control centers, still rely largely on centralized protocols. Without well defined standards for message formats, identity management, and trust anchors, blockchain-enabled AIVs risk becoming isolated islands of innovation unable to interact effectively with legacy systems. Compounding this concern is the looming quantum threat, as advances in quantum computing may render widely adopted cryptographic schemes obsolete, exposing vehicular blockchains to unprecedented vulnerabilities.

Looking forward, these challenges chart a clear agenda for future research. Lightweight, quantum-resistant cryptographic schemes tailored for embedded vehicle hardware will be crucial for long-term resilience. In parallel, adaptive consensus mechanisms that dynamically adjust to context, employing lightweight local validation for urgent safety events while reserving robust, global consensus for long-term accountability, offer a pathway to reconciling latency and security. Equally important are multi-layer blockchain architectures, where local, regional, and global tiers distribute responsibility and prevent systemic congestion. As 6G networks emerge, the integration of blockchain with ultra-low-latency, high-bandwidth communication and edge computing promises to reduce propagation delays and improve reliability. Finally, combining blockchain with AI-driven intelligence through anomaly detection, validator prediction, or federated learning, may yield systems that are not only more adaptive but also more transparent and auditable. In summary, addressing the blockchain trilemma in AIVs requires more than incremental optimization. It demands a multidisciplinary effort that unites advances in cryptography, distributed systems, vehicular communication, and artificial intelligence. Only through such integration can blockchain evolve from an experimental framework to a cornerstone of secure, scalable, and future-proof intelligent mobility. Table 5 outlines the challenges of the blockchain trilemma and links them to corresponding future research directions. It shows how issues in security, speed, scalability, and energy efficiency ranging from hardware limits and consensus delays to high power consumption can be addressed through innovations such as quantum-resistant cryptography, adaptive consensus, hierarchical blockchain architectures, and energy-aware protocols.

7 Conclusion

This chapter has critically examined the blockchain trilemma in the context of AIVs, showing that the interplay between security, speed, and scalability cannot be resolved through conventional blockchain architectures alone. Although blockchain holds immense promise for enhancing trust, integrity, and accountability in vehicular networks, the critical safety requirements of AIVs amplify the constraints of the trilemma. Through structured analysis, we demonstrated how secure sensor fusion, node authentication, latency-sensitive consensus, and scalable ledger designs interact within AIV ecosystems. We further illustrated, through case studies of dense urban intersections and highway platooning, how blockchain-based frameworks must adapt dynamically to the unique demands of traffic density and high-speed maneuvering. The proposed innovations, including hybrid consensus mechanisms, edge-enabled blockchain layers, AI-augmented security modules, and lightweight cryptographic primitives, suggest that the trilemma can be managed, though not eliminated, through careful system design. Importantly, the chapter underscored that solving this challenge requires a multi-disciplinary approach that integrates advances in cryptography, distributed systems, vehicular communication, and artificial intelligence.

Table 5 Challenges of the blockchain trilemma in AIVs and future research directions

Trilemma dimension	Current challenges in AIVs	Future research directions
Security	Hardware limits restrict cryptographic processing on embedded vehicle systems. Latency–security trade-off: stronger validation slows real-time decisions. Vulnerability to Sybil, replay, and quantum-based attacks.	Lightweight, quantum-resistant cryptography for constrained devices. AI-driven validator selection and anomaly detection to strengthen trust. Secure interoperability frameworks with legacy infrastructure
Speed (Latency)	Consensus delays threaten safety-critical actions (e.g., braking). Block propagation slows in dense networks.	Adaptive/dynamic consensus protocols (switch between lightweight local and robust global). Edge-integrated blockchain with 6G for ultra-low latency. Layer-2 offloading (state channels, sidechains) for rapid transactions
Scalability	Urban networks generate massive V2V/V2I events. Storage and bandwidth requirements exceed vehicle capabilities.	Hierarchical blockchain layers (local, regional, global). Sharding and cluster-based consensus to distribute load. Data minimization and event prioritization strategies
Energy Efficiency	High consensus/crypto overhead drains EV batteries. Communication-intensive protocols increase power use.	Energy-aware consensus mechanisms balancing load vs security. Scheduling validation during charging or off-peak use. Optimizing cryptography for low-power hardware

Looking ahead, research on quantum-resistant security schemes, integration with ultra-low-latency 6G networks, and decentralized AI-driven governance will be pivotal. The blockchain trilemma in AIVs should not be seen as an insurmountable limitation, but as a guiding principle to balance trade-offs while designing resilient, transparent, and future-proof vehicular ecosystems. In conclusion, this chapter establishes how blockchain trilemma research guides the design of secure AIV ecosystems. By situating security, speed, and scalability as interdependent pillars, the chapter complements broader discussions in this book on blockchain-based vehicular security and connectivity. The proposed innovations demonstrate practical pathways for deploying blockchain in mobility systems, while the identified research agenda, quantum-safe cryptography, AI-driven consensus, and 6G integration points toward future intelligent transportation frameworks.

References

1. Willie, A.: The role of blockchain in enhancing security for autonomous vehicles, Stanford University (2025). Ph.D. thesis
2. Nakai, T., Sakurai, A., Hironaka, S., Shudo, K.: A formulation of the trilemma in proof of work blockchain. IEEE Access **12**, 80559–80578 (2024)
3. Mssassi, S., Abou El Kalam, A.: The blockchain trilemma: A formal proof of the inherent trade-offs among decentralization, security, and scalability. Appl. Sci. **15**(1), 19 (2024)
4. Bendiab, G., Hameurlaine, A., Germanos, G., Kolokotronis, N., Shiaeles, S.: Autonomous vehicles security: challenges and solutions using blockchain and artificial intelligence. IEEE Trans. Intell. Transp. Syst. **24**(4), 3614–3637 (2023)
5. Iordache, S., Patilea, C.C., Paduraru, C.: Enhancing autonomous vehicle safety with blockchain technology: securing vehicle communication and AI systems. Future Internet **16**(12), 471 (2024)
6. Kakkar, R., Gupta, R., Agrawal, S., Tanwar, S., Sharma, R.: Blockchain-based secure and trusted data sharing scheme for autonomous vehicle underlying 5g. J. Inf. Sec. Appl. **67**, 103179 (2022)
7. Ajakwe, S.O., Olabisi, K.L., Kim, D.-S.: Multihop intruder node detection scheme (minds) for secured drones' fanet communication. IET Intel. Transport Syst. **19**(1), 70080 (2025)
8. Ajakwe, S.O., Igboanusi, I.S., Lee, J.-M., Kim, D.-S.: i banda: A blockchain-assisted defense system for authentication in drone-based logistics. Drones **9**(8), 590 (2025)
9. Chai, H., Leng, S., Wu, F., He, J.: Secure and efficient blockchain-based knowledge sharing for intelligent connected vehicles. IEEE Trans. Intell. Transp. Syst. **23**(9), 14620–14631 (2021)
10. Sun, J., Shi, Q., Jin, G., Xu, H., Chen, W.: Blockchain-enabled iov: Secure communication and trustworthy decision-making. In: 2024 IEEE Conference on Dependable and Secure Computing (DSC), pp. 104–105. IEEE (2024)
11. Ajakwe, S.O., Kim, D.-S., Lee, J.-M.: Drone transportation system: systematic review of security dynamics for smart mobility. IEEE Internet Things J. **10**(16), 14462–14482 (2023)
12. Reiff, N.: Scalability: Blockchain tech's greatest problem. Technical report, Investopedia (2024)
13. Ajakwe, S.O., Kim, D.-S.: Facets of security and safety problems and paradigms for smart aerial mobility and intelligent logistics. IET Intel. Transp. Syst. **18**, 2827–2855 (2024)
14. Monte, G.D., Pennino, D., Pizzonia, M.: Scaling blockchains without giving up decentralization and security: A solution to the blockchain scalability trilemma. In: Proceedings of the 3rd Workshop on Cryptocurrencies and Blockchains for Distributed Systems, pp. 71–76. (2020)
15. Ajakwe, S.O., Okoloegbo, C.A., Lee, J.M., Kim, D.S.: Machine learning models for drone security: Cognitive versus cyber intelligence for safety operations. In: Machine Learning for Drone-Enabled IoT Networks: Opportunities. Developments, and Trends, pp. 121–139. Springer, Cham (2025)
16. Li, S., Yu, M., Yang, C.-S., Avestimehr, A.S., Kannan, S., Viswanath, P.: Polyshard: Coded sharding achieves linearly scaling efficiency and security simultaneously. IEEE Trans. Inf. Forensics Secur. **16**, 249–261 (2020)
17. Lee, C.B.: Maimai: What is the blockchain trilemma and how to solve it? Technical report, One Key, Moon Pay (2025)
18. Quattrocchi, G., Scaramuzza, F., Tamburri, D.A.: The blockchain trilemma, (2024)
19. López-Ramírez, J.L., Jimenez-Mendoza, D., Benitez-Quintero, J.M., Avina-Bravo, E.G., Asael, G.-H.D., Gabriel, A.-C.J., et al.: An integrated blockchain framework for secure autonomous vehicle communication system. Information **16**(7), 557 (2025)
20. Al-Kafi, G.A., Ali, G., Reno, S.: Rewriting blockchain: a hybrid consensus that defeats the trilemma paradox. Comput. Electr. Eng. **126**, 110494 (2025)
21. Reno, S., Roy, K.: Navigating the blockchain trilemma: A review of recent advances and emerging solutions in decentralization, security, and scalability optimization. Comput. Mater. Continua 84(2), (2025)
22. Durlik, I., Miller, T., Kostecka, E., Zwierzewicz, Z., Łobodzińska, A.: Cybersecurity in autonomous vehicles–are we ready for the challenge? Electronics **13**(13), 2654 (2024)

23. Sahoo, S.: Sensor fusion and virtual sensor design for enhanced multi-sensor data accuracy in autonomous systems. Int. J. Smart & Sustain. Intell. Comput. **1**(2), 21–39 (2024)
24. Ogunbunmi, S., Aluko, S., Onukak, P.: Cyber-resilient perception: safeguarding autonomous vehicles with trust-aware sensor fusion. IEEE Sensors Rev , (2025)
25. Fan, Z., Zhao, L.: Autonomous integrity of multi-source information resilient fusion navigation system: state-of-the-art and open challenges. IEEE Trans. Instrum, Meas (2024)
26. Sun, M.: Multi-sensor data fusion and management strategies for robust perception in autonomous vehicles. Nuvern Appl. Sci. Rev. **8**(10), 59–68 (2024)
27. Kumari, A., Kumar, S., Raw, R.S.: Investigating the reliability of vehicle-to-vehicle and vehicle-to-infrastructure communication. In: International Conference on Innovations in Computational Intelligence and Computer Vision, pp. 483–500. Springer (2024)
28. Du, Q., Zhou, J., Ma, M.: Eaia: An efficient and anonymous identity authentication protocol for v2v communications in internet of vehicles. In: 2024 IEEE World AI IoT Congress (AIIoT), pp. 591–597. IEEE (2024)
29. Muhammad, M.: Efficient authentication of vehicle-to-vehicle (v2v) safety messages, Birmingham City University (2024). PhD thesis
30. Diego, S., Regueiro, C., Maciá-Fernández, G.: An authentication system based on self-sovereign identity for vehicle-to-vehicle (v2v) communications. In: International Congress on Blockchain and Applications, pp. 13–22. Springer (2024)
31. Sharma, A., Rani, S.: Context-aware authentication framework for secure v2v and v2i communications in autonomous vehicles using llm. IEEE Trans. Intell. Transp. Syst. , 1–8 (2025)
32. Ghafoor, M.I., Naeem, A.B., Senapati, B., Sudman, M.S.I., Pradhan, S., Das, D., Almeida, F., Sakr, H.A.: Privacy-preserving and lightweight v2i and v2v authentication protocol using blockchain technology. Intell. Autom. Soft Comput 39(5), (2024)
33. Abreu, R., Branco, F., Reis, M.J.C.S., Serôdio, C.: Cybersecurity in connected and autonomous vehicles: a systematic review of automotive security. IEEE Access **13**, 116818–116855 (2025)
34. Hemani, Singh, D., Dwivedi, R.K.: Designing blockchain based secure autonomous vehicular internet of things (iot) architecture with efficient smart contracts. Int. J. Inf. Technol. **17**(2), 1207–1223 (2025)
35. Lippi, G., Aljawarneh, M., Al-Na'amneh, Q., Hazaymih, R., Dhomeja, L.D., et al.: Security and privacy challenges and solutions in autonomous driving systems: a comprehensive review. J. Cyber Sec. Risk Auditing **2025**(3), 23–41 (2025)
36. Beni Prathiba, S., Murali, P., Shenbaga Moorthy, R., Kumar Anandhan, D., Selvaraj, A.K., Rodrigues, J.J.P.C.: A blockchain-powered malicious node detection in internet of autonomous vehicles. IEEE Trans. Intell. Transp. Syst. **25**(11), 16277–16287 (2024)
37. Hussain, S., Tahir, S., Masood, A., Tahir, H.: Blockchain-enabled secure communication framework for enhancing trust and access control in the internet of vehicles (iov). IEEE Access **12**, 110992–111006 (2024)
38. Zhang, C., Ji, M., Wu, Q., Fan, P., Fan, Q.: Optimizing system latency for blockchain-encrypted edge computing in internet of vehicles (2025). arXiv:2506.14208
39. Gao, F., Yang, B., Chen, C., Guan, X.: Edge-assisted flexible platooning of connected and automated vehicles under traffic anomaly. IEEE Trans. Veh. Technol. **74**(7), 10198–10208 (2025)
40. Thornton, S., Santhanam, N., Chhajer, R., Dey, S.: Real-time heterogeneous collaborative perception in edge-enabled vehicular environments. IEEE Open J. Veh. Technol. **6**, 471–486 (2025)
41. Jain, A.K., Gupta, N., Gupta, B.B.: A survey on scalable consensus algorithms for blockchain technology. Cyber Sec. Appl. **3**, 100065 (2025)
42. Luo, H., Zhang, J., Li, X., Li, Z., Yu, H., Sun, G., Niyato, D.: Esia: an efficient and stable identity authentication for internet of vehicles. IEEE Trans. Veh. Technol. **73**(4), 5602–5615 (2023)
43. Alghamdi, T.A., Khalid, R., Javaid, N.: A survey of blockchain based systems: Scalability issues and solutions, applications and future challenges. IEEE Access **12**, 79626–79651 (2024)

44. Liu, X., Xiong, X., Wang, J., Qi, Y.: An internet of vehicles road traffic data sharing scheme based on signcryption and editable blockchain. Comput. Netw. **255**, 110864 (2024)
45. Jatoth, C., Doriya, R., et al.: Iov block secure: blockchain based secure data collection and validation framework for internet of vehicles network. Peer-to-Peer Netw. Appl. **17**(6), 3964–3990 (2024)
46. Zhang, L., Hang, L., Zu, K., Wang, Y., Yang, K.: Dynamic vehicle reputation consensus: Enhancing iov communication with a blockchain algorithm. IEEE Trans. Veh. Technol. **74**(3), 4788–4806 (2025)
47. Grosset, J.: Collective intelligence strategies for efficient autonomous industrial vehicles, Ecole nationale supérieure Mines-Télécom Atlantique (2024). Ph.D. thesis
48. Al-awamy, A.A., Al-shaibany, N., Sikora, A., Welte, D.: Hybrid consensus mechanisms in blockchain: A comprehensive review. Int. J. Intell. Syst. **2025**(1), 5821997 (2025)
49. Wahab, N.A., Zhang, D., Juniardi, N., Wong, Y.: Advances in consortium chain scalability: A review of the practical byzantine fault tolerance consensus algorithm. Int. J. Adv. Comput. Sci. Appl. (IJACSA) , 7 (2024)
50. Deshmukh, A., Baiju, D., Atish, A., Alladi, T., Yu, F.R.: An efficient and scalable byzantine fault tolerant consensus for vehicular networks. IEEE Trans. Veh. Technol. **74**(9), 1–12 (2025)
51. Li, R.: Comparative analysis and future directions of consensus algorithms in blockchain technology. In: 2024 2nd International Conference on Image, Algorithms and Artificial Intelligence (ICIAAI 2024), pp. 333–344 (2024). Atlantis Press
52. Shafin, K.M., Reno, S.: A hybrid blockchain-enabled payment system for efficient electronic vehicle charging payments on electrified roads. J. King Saud Univ. Comput. Inf. Sci **37**(3), 1–21 (2025)
53. Venkatesan, K., Rahayu, S.B.: Blockchain security enhancement: an approach towards hybrid consensus algorithms and machine learning techniques. Sci. Rep. **14**(1), 1149 (2024)
54. Abinivesh, S., Isravel, D.P., Dhas, J.P.M.: Quorum-based tier approach: Integrating pbft and pos in a hybrid consensus mechanism. In: 2024 3rd International Conference on Automation, Computing and Renewable Systems (ICACRS), pp. 147–154. IEEE (2024)
55. Thangam, S., Chakkaravarthy, S.S.: An edge enabled region-oriented dag-based distributed ledger system for secure v2x communication. KSII Trans. Internet & Inf. Syst. **18**(8) (2024)
56. Ahmadvand, H., Foroutan, F.: Latency and privacy-aware resource allocation in vehicular edge computing, (2025). arXiv:2501.02804
57. Rathore, R., Kaushik, P., Sikarwar, S.S., Joshi, H., Mishra, A.K., Hudda, Y.: Intelligent transportation systems make use of fog and edge computing for navigation. In: 2024 IEEE International Conference on Interdisciplinary Approaches in Technology and Management for Social Innovation (IATMSI), vol. 2, pp. 1–6. (2024)
58. Badidi, E., El Harrouss, O.: Integrating edge computing and blockchain for safer and more efficient digital transportation systems. Proc. Comput. Sci. **251**, 273–280 (2024)
59. Khan, A.A., Laghari, A.A., Baqasah, A.M., Alroobaea, R., Almadhor, A., Sampedro, G.A., Kryvinska, N.: Blockchain-enabled infrastructural security solution for serverless consortium fog and edge computing. PeerJ Comput. Sci. **10**, 1933 (2024)
60. Zang, H., Kim, H., Kim, J.: Blockchain-based decentralized storage design for data confidence over cloud-native edge infrastructure. IEEE Access **12**, 50083–50099 (2024)
61. Abou El Houda, Z., Moudoud, H., Brik, B., Khoukhi, L.: Blockchain-enabled federated learning for enhanced collaborative intrusion detection in vehicular edge computing. IEEE Trans. Intell. Transp. Syst. **25**(7), 7661–7672 (2024)
62. Shivhare, A., Raw, R.S.: Vanet: Impact of blockchain and artificial intelligence on autonomous vehicles. In: 2025 International Conference on Networks and Cryptology (NETCRYPT), pp. 109–114. (2025)
63. G Anand, L.: Decentralized intrusion detection in vanets using blockchain and deep learning. Decentralized Intrusion Detection in VANETs Using Blockchain and Deep Learning (March 25, 2025) (2025)
64. Jetani, H., Patel, J., Mahida, N., Patel, M., Gupta, R., Tanwar, S., Gupta, A.: Blockchain and deep learning-based decentralized anomaly detection framework for vanet. In: 2024 IEEE International Conference on Vehicular Electronics and Safety (ICVES), pp. 1–6 (2024). IEEE

65. Mansouri, F., Tarhouni, M., Alaya, B., Zidi, S.: A distributed intrusion detection framework for vehicular ad hoc networks via federated learning and blockchain. Ad Hoc Netw. **167**, 103677 (2025)
66. Deng, Z., Tang, C., Li, T., He, D.: Permissioned blockchain-based trusted and robust consensus optimization orienting intelligent transportation systems. IEEE Trans. Intell. Transp. Syst. **26**(2), 2011–2026 (2025)
67. Bijalwan, J.G., Singh, J., Ravi, V., Bijalwan, A., Alahmadi, T.J., Singh, P., Diwakar, M.: Navigating the future of secure and efficient intelligent transportation systems using AI and blockchain. The Open Transp. J. **18**(1) (2024)
68. Zheng, G., Ni, Q., Lu, Y.: Privacy-aware anomaly detection and notification enhancement for vanet based on collaborative intrusion detection system. IEEE Trans. Intell. Transp. Syst. **25**(12), 21172–21182 (2024)
69. Rajbhar, S.K., Singh, S.: A comprehensive survey of lightweight authentication mechanisms for vehicle group communication in 5g networks. In: 2025 IEEE International Students' Conference on Electrical, Electronics and Computer Science (SCEECS), pp. 1–7. (2025)
70. Ahmed, B., Zakarya, B.: Anel: a novel efficient and lightweight authentication scheme for enhancing security in vehicular ad hoc networks (vanets) using elliptic curve cryptography. Stud. Eng. Exact Sci. **5**(2), 9180–9180 (2024)
71. Castiglione, A., Elia, T.: Enhancing vehicle communication security: Implementing post-quantum solutions for can networks. In: International Conference on Advanced Information Networking and Applications, pp. 354–366. Springer (2025)
72. Aslam, A.M., Bhardwaj, A., Chaudhary, R.: Quantum-resilient blockchain-enabled secure communication framework for connected autonomous vehicles using post-quantum cryptography. Veh. Commun. **52**, 100880 (2025)
73. Yadav, D.K., Sahu, H., Chaudhary, D., Lee, C.-C.: Module lattice based post quantum secure blockchain empowered vehicle to vehicle communication in the internet of vehicles. Comput. Electr. Eng. **117**, 109245 (2024)
74. Castiglione, A., Elia, T.: Securing in-vehicle communications through post-quantum cryptography and network segmentation. Comput. Electr. Eng. **126**, 110488 (2025)
75. AlMarshoud, M., Sabir Kiraz, M., H. Al-Bayatti, A.: Security, privacy, and decentralized trust management in vanets: A review of current research and future directions. ACM Comput. Surv. **56**(10), 1–39 (2024)
76. Ullah, I., Deng, X., Pei, X., Mushtaq, H., Khan, Z.: Securing internet of vehicles: a blockchain-based federated learning approach for enhanced intrusion detection. Clust. Comput. **28**(4), 256 (2025)
77. Liu, P., He, Q., Chen, Y., Jiang, S., Zhao, B., Wang, X.: A lightweight authentication and privacy-preserving aggregation for blockchain-enabled federated learning in vanets. IEEE Trans. Consum. Electron. **71**(1), 1274–1287 (2025)
78. Shen, X., Ma, R.: A blockchain solution for the internet of vehicles with better filtering and adaptive capabilities. Sensors **25**(4), 1030 (2025)
79. Decker, N.: Natra: blockchain-based national traffic architecture for real-time routing, jurisdictional equity, and autonomous vehicle interoperability. Jurisdictional Equity, and Autonomous Vehicle Interoperability: (June 21, 2025) (2025)
80. George, D., Pavithra, S., Das, J.: Cyber-resilient autonomous vehicles: Securing networks and enhancing decision-making with next-gen security measures. Results Eng , 107179 (2025)
81. Singh Rawat, G., Singh, K., Shariq, M., Das, A.K., Ashraf Chaudhry, S., Lorenz, P.: Btc2pa: a blockchain-assisted trust computation with conditional privacy- preserving authentication for connected vehicles. IEEE Trans. Intell. Transp. Syst. **26**(1), 1134–1148 (2025)
82. Badshah, A., Abbas, G., Waqas, M., Muhammad, F., Abbas, Z.H., Bilal, M., Song, H.: Blockchain-assisted lightweight authenticated key agreement security framework for smart vehicles-enabled intelligent transportation system. IEEE Trans. Autom. Sci. Eng. **21**(3), 2425–2439 (2024)
83. Surapaneni, P., Bojjagani, S., Khurram Khan, M.: Dynamic-trust: Blockchain-enhanced trust for secure vehicle transitions in intelligent transport systems. IEEE Trans. Intell. Transp. Syst. **26**(7), 10918–10932 (2025)

84. Zhao, W., Luo, X.: Layer 2 blockchains: An introduction. In: Proceedings of the 2024 7th Artificial Intelligence and Cloud Computing Conference, pp. 449–456. (2024)
85. Vo, K.T., Nguyen, H.-T., Nguyen-Hoang, T.-A.: An integrated decision-making framework for enhancing blockchain scalability and efficiency: Layer-2, off-chain storage, and smart contracts. IEEE Access (2025)
86. Quan, B.L.Y., Wahab, N.H.A., Al-Dhaqm, A., Alshammari, A., Aqarni, A., Abd Razak, S., Wei, K.T.: Recent advances in sharding techniques for scalable blockchain networks: A review, IEEE Access (2024)
87. Huang, J., Kong, L., Wang, J., Chen, G., Gao, J., Huang, G., Khan, M.K.: Secure data sharing over vehicular networks based on multi-sharding blockchain. ACM Trans. Sensor Netw. **20**(2), 1–23 (2024)
88. Murugan, K.S., Lakshmi, N.A., Priya, N.S.: Ai-driven blockchain consensus for efficient node selection in cluster-based vehicular networks. In: 2024 International Conference on Innovative Computing, Intelligent Communication and Smart Electrical Systems (ICSES), pp. 1–7. IEEE (2024)
89. Chai, H., Leng, S., Chen, Y., Zhang, K.: A hierarchical blockchain-enabled federated learning algorithm for knowledge sharing in internet of vehicles. IEEE Trans. Intell. Transp. Syst. **22**(7), 3975–3986 (2021)
90. Boujezza, H., Boubakri, A.: A new redundant intelligent architecture to improve the operational safety of autonomous vehicles. In: International Conference on Advanced Information Networking and Applications, pp. 140–152. Springer (2024)
91. Mahmood, A., Szabolcsi, R.: A systematic review on risk management and enhancing reliability in autonomous vehicles. Machines **13**(8), 646 (2025)
92. Sun, Z.: Fault-tolerant control of intelligent transportation vehicles based on instant learning and heuristic dynamic planning. Int. J. Reasoning-based Intell. Syst. **17**(8), 21–28 (2025)
93. Srivastava, S., Agarwal, D., Chaurasia, B.K., Adhikari, M.: Blockchain-based trust management for data exchange in internet of vehicle network. Multimed. Tools Appl. **84**(8), 4837–4855 (2025)
94. Zhu, P., Ferrari-Trecate, G., Geroliminis, N.: Hierarchical control for vehicle repositioning in autonomous mobility-on-demand systems. IEEE Trans. Control Syst. Technol. **33**(4), 1463–1476 (2025)
95. Fardad, M., Muntean, G.-M., Tal, I.: A blockchain-enabled vehicular edge computing framework for secure performance-oriented v2x service delivery. IEEE Trans. Veh. Technol. **73**(9), 13853–13867 (2024)
96. Liu, Y., Liu, A., Lu, Y., Pan, Z., Li, Y., Liu, J., Bian, S., Conti, M.: Kronos: A secure and generic sharding blockchain consensus with optimized overhead. Cryptology ePrint Archive (2024)
97. Hong, Z., Guo, S., Zhou, E., Chen, W., Huang, H., Zomaya, A.: Gridb: Scaling blockchain database via sharding and off-chain cross-shard mechanism, (2024). arXiv:2407.03750
98. Yang, N., Tang, C., Zong, T., Zeng, Z., Xiong, Z., He, D.: Ric-sda: A reputation incentive committee-based secure conditional dual authentication scheme for vanets. IEEE Trans, Mobile Comput (2024)
99. Mohiuddin, M.A., Nirosha, K., Anusha, D., Nazeer, M., NV, G.R., Lakhanpal, S.: Ai to v2x privacy and security issues in autonomous vehicles: survey. In: MATEC Web of Conferences, vol. 392, p. 01097 (2024). EDP Sciences
100. Hakeem, S.A.A., Kim, H.: Advancing intrusion detection in v2x networks: A comprehensive survey on machine learning, federated learning, and edge AI for v2x security. IEEE Trans. Intell. Transp, Syst (2025)
101. Protim, J., Sharmin, Z., Sultana, B.: Hierarchical federated learning for internet of vehicles networks: A multi-tiered approach to efficient and adaptive distributed learning. In: 2024 6th International Conference on Sustainable Technologies for Industry 5.0 (STI), pp. 1–6. IEEE (2024)
102. Bentayeb, Y., Chaoui, K., Badir, H.: Integrating blockchain and edge computing: A systematic analysis of security, efficiency, and scalability. Int. J. Adv. Comput. Sci. Appl. 16(1), (2025)

The manufacturer's authorised representative in the EU is Springer Nature Customer Service Centre GmbH, Europaplatz 3, 69115 Heidelberg, Germany. If you have any concerns regarding our products, please contact ProductSafety@springernature.com

Printed and bound by CPI Group (UK) Ltd, Croydon, CR0 4YY
05/10/2026
02228057-0003